Carbon-Based Conductive Polymer Composites

Carbon nanomaterials can transfer their excellent electrical conductivity to polymers while enhancing or maintaining their original mechanical properties. Conductive polymer composites based on carbon nanomaterials are finding increasing applications in aerospace, automotive, and electronic industries when flexibility or lightweight is required. *Carbon-Based Conductive Polymer Composites: Processing, Properties, and Applications in Flexible Strain Sensors* summarizes recent remarkable achievements in the processing–structure–property relationship of conductive polymer composites based on carbon nanomaterials. It also discusses research developments for their application in flexible strain sensors and novel processing methods like additive manufacturing.

- Presents the state of the art in conductive composite materials and their application in flexible strain sensors.
- Uniquely combines the processing, structure, properties, and applications of conductive polymer composites.
- Integrates theory and practice.
- Benefits plastics converters who wish to take full advantage of the potential of conductive plastic materials.

This book is written for material scientists and engineers researching and applying these advanced materials for a variety of applications.

Emerging Materials and Technologies
Series Editor: *Boris I. Kharissov*

The *Emerging Materials and Technologies* series is devoted to highlighting publications centered on emerging advanced materials and novel technologies. Attention is paid to those newly discovered or applied materials with potential to solve pressing societal problems and improve quality of life, corresponding to environmental protection, medicine, communications, energy, transportation, advanced manufacturing, and related areas.

The series takes into account that, under present strong demands for energy, material, and cost savings, as well as heavy contamination problems and worldwide pandemic conditions, the area of emerging materials and related scalable technologies is a highly interdisciplinary field, with the need for researchers, professionals, and academics across the spectrum of engineering and technological disciplines. The main objective of this book series is to attract more attention to these materials and technologies and invite conversation among the international R&D community.

Advanced Materials for a Sustainable Environment: Development Strategies and Applications
Edited by Naveen Kumar and Peter Ramashadi Makgwane

Nanomaterials from Renewable Resources for Emerging Applications
Edited by Sandeep S. Ahankari, Amar K. Mohanty, and Manjusri Misra

Multifunctional Polymeric Foams: Advancements and Innovative Approaches
Edited by Soney C George and Resmi B.P.

Nanotechnology Platforms for Antiviral Challenges: Fundamentals, Applications and Advances
Edited by Soney C George and Ann Rose Abraham

Carbon-Based Conductive Polymer Composites: Processing, Properties, and Applications in Flexible Strain Sensors
Dong Xiang

Nanocarbons: Preparation, Assessments, and Applications
Ashwini P. Alegaonkar and Prashant S. Alegaonkar

Emerging Applications of Carbon Nanotubes and Graphene
Edited by Bhanu Pratap Singh and Kiran M. Subhedar

For more information about this series, please visit: www.routledge.com/Emerging-Materials-and-Technologies/book-series/CRCEMT

Carbon-Based Conductive Polymer Composites
Processing, Properties, and Applications in Flexible Strain Sensors

Dong Xiang

CRC Press is an imprint of the
Taylor & Francis Group, an **informa** business

First edition published 2023
by CRC Press
6000 Broken Sound Parkway NW, Suite 300, Boca Raton, FL 33487-2742

and by CRC Press
4 Park Square, Milton Park, Abingdon, Oxon, OX14 4RN

CRC Press is an imprint of Taylor & Francis Group, LLC

© 2023 Dong Xiang

Reasonable efforts have been made to publish reliable data and information, but the author and publisher cannot assume responsibility for the validity of all materials or the consequences of their use. The authors and publishers have attempted to trace the copyright holders of all material reproduced in this publication and apologize to copyright holders if permission to publish in this form has not been obtained. If any copyright material has not been acknowledged please write and let us know so we may rectify in any future reprint.

Except as permitted under U.S. Copyright Law, no part of this book may be reprinted, reproduced, transmitted, or utilized in any form by any electronic, mechanical, or other means, now known or hereafter invented, including photocopying, microfilming, and recording, or in any information storage or retrieval system, without written permission from the publishers.

For permission to photocopy or use material electronically from this work, access www.copyright.com or contact the Copyright Clearance Center, Inc. (CCC), 222 Rosewood Drive, Danvers, MA 01923, 978-750-8400. For works that are not available on CCC please contact mpkbookspermissions@tandf.co.uk

Trademark notice: Product or corporate names may be trademarks or registered trademarks and are used only for identification and explanation without intent to infringe.

ISBN: 978-1-032-11158-2 (hbk)
ISBN: 978-1-032-11159-9 (pbk)
ISBN: 978-1-003-21866-1 (ebk)

DOI: 10.1201/9781003218661

Typeset in Times
by codeMantra

Contents

Preface xi
Author xiii

Chapter 1   Introduction 1

    1.1   Introduction 1
    1.2   Preparation of Conductive Polymer Composites 2
            1.2.1   In situ Polymerization 2
            1.2.2   Solution Mixing 2
            1.2.3   Melt Mixing 2
    1.3   Processing of Conductive Polymer Composites 3
            1.3.1   Compression Molding 3
            1.3.2   Biaxial Stretching 3
            1.3.3   Blown Film Extrusion 3
            1.3.4   Injection Molding 4
            1.3.5   Casting 4
            1.3.6   Other Processing Methods 4
    1.4   Properties of Conductive Polymer Composites 4
            1.4.1   Mechanical Properties 5
            1.4.2   Electrical Properties 5
            1.4.3   Thermal Properties 5
            1.4.4   Barrier Properties 5
    1.5   Specially Designed Structure 6
            1.5.1   Segregated Structure 7
            1.5.2   Double Percolated Structure 7
    1.6   Applications of Conductive Polymer Composites in
         Flexible Strain Sensors 7
            1.6.1   Fabrication of Flexible Strain Sensors 8
                    1.6.1.1   Coating 8
                    1.6.1.2   Electrostatic Self-Assembly 8
                    1.6.1.3   3D Printing 9
                    1.6.1.4   Chemical Vapor Deposition 9
                    1.6.1.5   Other Methods 9
            1.6.2   Sensing Mechanism of Flexible Strain Sensors 10
                    1.6.2.1   Tunneling Effect 10
                    1.6.2.2   Geometric Effect 10
                    1.6.2.3   Piezoresistive Effect 11
                    1.6.2.4   Crack Propagation 12
                    1.6.2.5   Disconnection Mechanism 12
            1.6.3   Sensing Performances 12
                    1.6.3.1   Strain Detection Range 12
                    1.6.3.2   Sensitivity 13

		1.6.3.3	Linearity 13
		1.6.3.4	Hysteresis 14
		1.6.3.5	Dynamic Durability 14
	1.7	Conclusions 14
	References 15

Chapter 2	Compression Molded Conductive Polymer Composites 19
	2.1	Introduction 19
	2.2	Compression Molded HDPE/MWCNT Composites 20
		2.2.1	Polarized Optical Microscopy 20
		2.2.2	Scanning Electron Microscopy 22
		2.2.3	Thermal Properties 23
		2.2.4	Tensile Properties 24
		2.2.5	Electrical Properties 26
	2.3	Unary Carbon Nanofiller-Reinforced Composites 29
		2.3.1	Scanning Electron Microscopy 29
		2.3.2	Thermal Properties 29
		2.3.3	Tensile Properties 30
		2.3.4	Electrical Properties 32
	2.4	Binary Carbon Nanofiller-Reinforced Composites 33
		2.4.1	Scanning Electron Microscopy 33
		2.4.2	Thermal Properties 34
		2.4.3	Tensile Properties 35
		2.4.4	Electrical Properties 37
	2.5	Conclusions 38
	References 39

Chapter 3	Biaxially Stretched Conductive Polymer Composites 43
	3.1	Introduction 43
	3.2	Biaxially Stretched HDPE/MWCNT Composites 44
		3.2.1	Biaxial Deformation Behavior 44
		3.2.2	Structural Evolution 46
		3.2.3	Thermal Properties 51
		3.2.4	Tensile Properties 52
		3.2.5	Electrical Properties 53
	3.3	Biaxially Stretched Unary Carbon Nanofiller-Reinforced Composites 56
		3.3.1	Biaxial Deformation Behavior 56
		3.3.2	Thermal Properties 57
		3.3.3	Structural Evolution 58
		3.3.4	Tensile Properties 60
		3.3.5	Electrical Properties 62
		3.3.6	Barrier Properties 63

Contents

	3.4	Biaxially Stretched Binary Carbon Nanofiller-Reinforced Composites 64
		3.4.1 Structural Evolution 64
		3.4.2 Thermal Properties 65
		3.4.3 Tensile Properties 65
		3.4.4 Electrical Properties 69
		3.4.5 Barrier Properties 70
	3.5	Biaxial Stretching of PP/MWCNT and TPU/MWCNT/rGO Composites 71
	3.6	Conclusions 74
	References 74
Chapter 4	Blown Film Extrusion of Conductive Polymer Composites 77
	4.1	Introduction 77
	4.2	Blown Film Extrusion of Thermoplastic Polyurethane/CNTs ... 78
		4.2.1 Morphology 78
		4.2.2 Structure 79
		4.2.3 Tensile Properties 80
		4.2.4 Dynamic-Mechanical Properties 81
	4.3	Blown Film Extrusion of TPU/Graphene Oxide Nanocomposites 81
		4.3.1 Thermal Properties 81
		4.3.2 Structure 82
		4.3.3 Barrier Properties 83
	4.4	Blown Film Extrusion of High-Density Polyethylene/CNT Composites 83
		4.4.1 Morphology 84
		4.4.2 Thermal Properties 84
		4.4.3 Tensile Properties 86
		4.4.4 Electrical Properties 88
	4.5	Blown Bubble Films of Aligned Nanowires and CNTs 90
		4.5.1 BBFs with Silicon Nanowires 90
		4.5.2 BBFs with CNTs 91
		4.5.3 BBFs with Large-Area Transistor Arrays 93
	4.6	Conclusions 94
	References 95
Chapter 5	Temperature-Resistivity and Damage Self-Sensing Behavior of Conductive Polymer Composites 97
	5.1	Introduction 97
	5.2	Construction of Conductive Network Structures in CPCs 98
	5.3	Stimuli-Resistivity Behaviors of CPCs 100
		5.3.1 Temperature-Resistivity Behavior 100
		5.3.2 Damage Self-Sensing Behavior 103

	5.4	Conclusions	108
	References		109

Chapter 6 Flexible Strain Sensors Based on Elastic Fibers of Conductive Polymer Composites 113

- 6.1 Introduction 113
- 6.2 Methods for Fabricating Flexible Fiber Strain Sensors 114
  - 6.2.1 LBL Coating and Ultrasonic-Assisted Dip-Coating... 114
  - 6.2.2 Chemical Deposition Coating 115
  - 6.2.3 Melt Extrusion 116
  - 6.2.4 Spinning 117
- 6.3 Relationship between Structure and Performance of Flexible Fiber Strain Sensor 118
  - 6.3.1 Sheath-Core Spun Yarn 118
  - 6.3.2 Helical Yarn 119
  - 6.3.3 Fabric 119
- 6.4 Applications of Flexible Fiber Strain Sensors 120
  - 6.4.1 Personal Health Care 120
  - 6.4.2 Body Motion Detection 120
  - 6.4.3 Human–Machine Interactions 121
  - 6.4.4 Intelligent Robotics 122
- 6.5 Conclusions 122
- References 123

Chapter 7 Flexible Strain Sensors Based on Sponges of Conductive Polymer Composites 127

- 7.1 Introduction 127
- 7.2 Types of Sponge-Based Strain Sensors 127
  - 7.2.1 Neat Conductive Sponge 127
  - 7.2.2 Conductive Sponge Impregnated with Elastomer 128
  - 7.2.3 Composite Conductive Sponge 128
  - 7.2.4 Conductive Material-Coated Sponge 128
- 7.3 Methods for Fabrication of Sponge-Based Strain Sensors 128
  - 7.3.1 Supercritical Foaming Technology 128
  - 7.3.2 Chemical Vapor Deposition (CVD) 129
  - 7.3.3 Freeze-Drying Method 129
  - 7.3.4 Template Method 131
    - 7.3.4.1 Carbonization of Template Sponge 131
    - 7.3.4.2 Template Removal Method 132
    - 7.3.4.3 Surface Coating of a Template Sponge 133
- 7.4 Application of Sponge-Based Strain Sensors 135
  - 7.4.1 Wearable Electronic Device 135
  - 7.4.2 Human–Computer Interaction/Intelligent Robot 136
  - 7.4.3 Electronic Skin 136

Contents

	7.5	Conclusions	137
		References	138

Chapter 8 3D-Printed Flexible Strain Sensors of Conductive Polymer Composites 141

- 8.1 Introduction 141
- 8.2 Preparation of 3D-Printed Strain Sensors 141
  - 8.2.1 DIW-Based 3D-Printed Strain Sensors 142
  - 8.2.2 SLA-Based 3D-Printed Strain Sensors 143
  - 8.2.3 SLS-Based 3D Printed Strain Sensors 143
  - 8.2.4 FDM-Based 3D-Printed Strain Sensors 144
- 8.3 Conductive Materials for 3D-Printed Strain Sensors 145
  - 8.3.1 Carbon Materials 145
  - 8.3.2 Metal Material/MXene 148
  - 8.3.3 Conductive Hydrogel 149
- 8.4 Architectural Design for 3D-Printed Strain Sensors 150
  - 8.4.1 Micro-Nano Porous Structure 150
  - 8.4.2 Bionic Structure 153
  - 8.4.3 Microstructure Channels 154
- 8.5 Application of 3D-Printed Strain Sensors 155
  - 8.5.1 Electronic Skin 155
  - 8.5.2 Soft Robotic Systems 156
  - 8.5.3 Wearable Electronic Devices 157
- 8.6 Conclusions 157

References 158

Index 161

Preface

Carbon-based nanomaterials, such as carbon nanotubes, graphene, and carbon black, are capable of transferring their excellent electrical conductivity to a polymer while also enhancing or maintaining its original mechanical properties. Conductive polymer composites based on carbon nanomaterials are finding increasing applications in aerospace, automotive and electronic industries for electrostatic dissipation, electromagnetic interference shielding, sensors, etc., when flexibility or lightweight is required. The influence of processing on the structuring and properties of these conductive polymer composites is usually ignored, yet this is a critical aspect of nanocomposites.

This book summarizes the recent remarkable achievements in the processing/structure/property relationship of conductive polymer composites based on carbon nanomaterials. This book also discusses the extensive research developments for their application in flexible strain sensors and novel processing methods like additive manufacturing. The book contains eight chapters, and each chapter addresses some specific issues related to conductive polymer composites based on carbon nanomaterials and also demonstrates the real potentialities of these materials in flexible strain sensors.

This book is suitable for researchers in the field of polymer composites and business managers. It can also be used as a reference book for undergraduate, graduate, and teachers majoring in materials. The key features of the book are as follows:

- Chapters written by the author in his own fields;
- Unique topic integration on the processing, structure, properties, and applications of conductive polymer composites;
- Presents the state of the art in conductive composite materials and their application in flexible strain sensors;
- Benefits plastics converters who wish to take full advantage of the potential of conductive plastic materials;
- Integrates theory and practice in the fields of conductive polymer composites.

My thanks are due to the team from CRC Press, Allison Shatkin, Hannah Warfel, and Gabrielle Vernachio for helping to bring this work to fruition, to my colleagues and postgraduates for their contributions to the chapters of this book, including Prof. Yuanpeng Wu, Prof. Yuntao Li, Dr. Chunxia Zhao, Dr. Hui Li, Prof. Bin Wang (Chapter 1), Zhouyu Liu, Zhen Li (Chapter 2), Xiaoyu Chen, Zhen Li, Ping Wang (Chapter 3), Jiayi Li, Zhenyu Li (Chapter 4), Lei Wang (Chapter 5), Qin Chen (Chapter 6), Xuezhong Zhang (Chapter 7), and Libing Liu (Chapter 8). Thanks to my family for the continuing support.

Dong Xiang
Chengdu, P.R. China

Author

Dong Xiang, PhD, is currently an associate professor in Southwest Petroleum University, China. He earned a PhD at Queen's University Belfast in 2015. His research interests mainly include conductive polymer composites, fiber-reinforced polymer composites, flexible strain sensors, and additive manufacturing. He has published more than 80 research articles in journals such as *Composites Science and Technology, Composites Part A: Applied Science and Manufacturing and Composites Part B: Engineering,* and *One Book*. He was awarded the Alan Glanvill Award by the Institute of Materials, Minerals and Mining (IOM3), and the Young Scientist Medal by International Association of Advanced Materials (IAAM) in 2022.

1 Introduction

1.1 INTRODUCTION

With the development of intelligent technologies, the use of wearable electronic devices in social life has gradually increased and became the backbone of intelligent electronics [1–4]. Wearable electronic devices made of conductive polymer composites (CPCs) have attracted wider attention owing to their high flexibility, superior stretchability, lightweight, low cost, and easy preparation [5–8]. CPCs are composed of conductive nanofillers and polymer matrices. Conductive nanofillers include carbon nanofillers and other nanofillers, in which carbon nanofillers mainly include materials like zero-dimensional carbon black (CB) [9]; one-dimensional carbon nanotubes (CNTs) [10,11]; two-dimensional graphene [12]; and its derivatives (graphite nanoplatelets (GNPs) [13,14], graphene oxide (GO) [15], and reduced graphene oxide (RGO) [16]). Also, polymer matrices mainly include polyethylene (PE) [17], polypropylene (PP) [18], polydimethylsiloxane (PDMS) [7,11,19], thermoplastic polyurethane (TPU) [20], and styrene butadiene styrene (SBS) [21,22]. In addition to electronic devices, CPCs are increasingly applied in aerospace, automotive, and electronic industries for anti-static protection, electrostatic dissipation, electromagnetic interference shielding, and flexible strain sensing [20,23–26]. In recent years, tremendous research has been carried out on the preparation and characterization of carbon-based CPCs [27,28]. The influence of preparation and processing on the internal conductive network structures and properties of CPCs has attracted less attention, while holding the key aspect in nanocomposites research. Studying the relationship between the internal structures, properties, and processing of CPCs is important for the development of CPCs. Likewise, predicting the internal structures and performances of CPCs through processing is an essential step for advanced applications of CPCs.

As the core of wearable electronic devices, flexible strain sensors have attracted increasing interest [4,5,29]. In addition to the characteristics of CPCs, flexible strain sensors are also characterized by good biocompatibility, small size, and diverse and low-cost preparation methods [30,31]. In this respect, Zhang et al. [32] developed a facile Dry-Meyer-Rod-Coating process to fabricate graphite/silk fiber strain sensors. However, the resulting sensors showed only 15% strain detection range and low sensitivity with a gauge factor (GF) of 14.5. Wang et al. [19] introduced CNTs to TPU fiber surface by ultrasonic treatment followed by encapsulation by PDMS to yield strain sensors with 100% strain monitoring range, but its maximum sensitivity is only 0.339. So far, realizing flexible strain sensors with wide strain detection ranges, high sensitivities, and other excellent sensing performances remains challenging. Therefore, searching for breakthrough directions for the development and application of flexible strain sensors with advanced characteristics is highly desirable.

In this chapter, the process/structure/property of carbon-based CPCs with different dispersion and orientation of carbon nanofillers prepared by methods like compression molding, biaxial stretching, blown film extrusion, injection molding, and

DOI: 10.1201/9781003218661-1

casting was addressed. The differences in properties were studied, and the relationships between the properties, structures, and processing of CPCs were clarified. In addition, the applications of carbon-based CPCs in flexible strain sensors were discussed. CPC-based flexible strain sensors with core-sheath structures, homogeneous structures, and segregated structures prepared by coating, electrostatic self-assembly, 3D printing, and chemical vapor deposition (CVD) were also presented. Moreover, the effects of non-covalent modifications, synergistic effects, and segregated structures on the sensing performance were analyzed in terms of the improved interaction force between nanofillers and polymer matrix, as well as the construction of complete and efficient percolation networks.

1.2 PREPARATION OF CONDUCTIVE POLYMER COMPOSITES

Three main methods are used to produce CPC nanofillers uniformly distributed and dispersed within polymer matrices: (i) in situ polymerization, (ii) solution mixing, and (iii) melt mixing.

1.2.1 IN SITU POLYMERIZATION

In in situ polymerization, nanofillers are first uniformly dispersed in monomer or monomer solution then followed by monomer polymerization in the presence of an initiator so that nanofillers are evenly dispersed in the polymer matrix to form CPCs. For instance, Wang et al. [33] prepared polypyrrole/silver composites by in situ polymerization. The sheet resistance of the composite was estimated to 61.54 Ω sq^{-1}. The dispersion of nanofillers within a polymer matrix can be improved by in situ polymerization, but the high cost of CPCs prepared by this method limits large-scale industrial production.

1.2.2 SOLUTION MIXING

In solution mixing, carbon nanofillers are first evenly dispersed in volatile organic solvents. The polymer matrix is then dissolved in organic solvents by mechanical mixing, magnetic stirring, or high-energy sonication. Finally, CPCs are obtained by evaporation of organic solvents. This method is effective for dispersing carbon nanofillers, but the toxicity of the organic solvents in terms of damage to the environment and the human body limits its use to small-scale preparation of CPCs for research purposes. Zheng et al. [9] prepared PDMS-based conductive nanocomposites by solution mixing and studied the effect of carbon nanofillers' dimensionality on the sensing performances of PDMS-based strain sensors. The results show that the CB/PDMS strain sensor (GF = 15.75) has higher sensitivity than the CNTs/PDMS strain sensor (GF = 4.36) at 10% strain.

1.2.3 MELT MIXING

Melt mixing can be used to disperse and mix the polymer matrix and carbon nanofillers above the polymer viscous flow temperature with the shear force generated by various mixing equipment for CPCs preparation. Su et al. [34] prepared CNT/

polycarbonate (PC)/poly(vinylidene fluoride) (PVDF) ternary blends by melt mixing. For PC: PVDF=3:7 and 5:5, the percolation thresholds reached 0.8 and 0.95 vol%, respectively. Compared to in situ polymerization and solution mixing, the dispersion of carbon nanofillers within the polymer matrix for CPCs prepared by melt mixing still requires improvement. Melt mixing can be used as a simple and effective method to prepare CPCs, especially for large-scale industrial production.

1.3 PROCESSING OF CONDUCTIVE POLYMER COMPOSITES

Polymers can be plasticized in a uniform viscous flow state suitable for molding into the desired shape and cured for shaping. The processing methods of polymers can be divided into thermoplasticization cooling molding, thermoplasticization reaction molding, and solvent plasticization dissolvent molding, including compression molding, biaxial stretching, blown film extrusion, injection molding, and casting. In general, the neat polymer processes might also be suitable for processing CPCs.

1.3.1 Compression Molding

Compression molding consists of directly adding powdery or loose granular solid plastics into a mold followed by gradual softening and melting by heating and pressurizing to form the desired shapes of the mold cavity before solidification into plastic parts. Compression molding is mainly used to form thermocuring plastics and thermoplastic plastics. CPCs might also be employed in compression molding. For example, CPCs can be granulated into particles, which may then be shaped by compression molding.

1.3.2 Biaxial Stretching

In biaxial stretching, the polymer is melted into a plasticized state in the extruder, which is extruded and produced by a flat die. The sheet is cooled above the glass transition temperature and kept hot before stretching in the transverse and longitudinal directions followed by cooling to yield biaxially stretched films. Biaxial stretching can be divided into sequential and simultaneous stretching. After biaxial stretching, the polymer chains are arranged parallel to the film plane to yield uniform strength of the film surface in all aspects, much higher than before stretching [35]. When biaxial stretching is applied to carbon-based CPCs, the interface interaction between the polymer matrix and carbon nanofillers would drive the carbon nanofillers by the movement of polymer chain, leading to improved dispersion of carbon nanofillers in the matrix.

1.3.3 Blown Film Extrusion

Blown film extrusion consists of heating and melting plastic followed by blowing to form a film. Usually, the polymer is extruded to form a tubular film blank. In a good melt flow state, the pipe film is blown to the required thickness through high-pressure air to become a film after cooling and shaping. Xiang et al. [17] prepared CNT/high-density polyethylene (HDPE) composites by blown film extrusion. The resulting blown film composites exhibited better mechanical properties thanks to the enhanced orientation and disentanglement of CNTs.

1.3.4 Injection Molding

The processing of CPCs by injection molding requires the addition of granular or powdered CPCs to the barrel of the injection molding machine. The mixture is then heated and melted to reach a flow state, followed by injection into the mold from the nozzle at the front end of the barrel under the rapid and continuous pressure of the plunger or moving screw of the injection machine. The melt filled with the mold under pressure is then cooled (thermoplastic) or heated (thermosetting plastic), and solidified. Finally, the mold is opened to obtain the product corresponding to the mold cavity.

1.3.5 Casting

Casting is a commonly used material processing technology that introduces the liquid CPCs or the suspension of CPCs into the mold for curing molding. Vadukumpully et al. [36] prepared graphene/polyvinyl chloride-based CPCs by casting to yield composites with high mechanical strength and thermal stability. CPCs obtained by solution casting are conducive to improve the dispersion of carbon nanofillers in polymer matrices. They are also convenient for large-scale industrial production due to their low cost [1].

1.3.6 Other Processing Methods

Other processing routes are also used for conductive polymer composites, including thermoforming and stretch blow molding. Thermoforming of conductive polymer composites requires CPCs sheets as raw materials. After heating, stretching, pressure, and cooling, the sheets are deformed and formed. Before using thermoforming, carrying out a sheet forming process for CPCs is required. After thermoforming, post-treatment with high processing cost is also needed. Stretch blow molding refers to a blow molding by biaxial directional stretching. The parison of CPCs is first formed by extrusion or injection. The resulting parison is then treated to the appropriate stretching temperature. Longitudinally stretching by internal (with stretching mandrel) or external (with stretching fixture) mechanical force and horizontally stretched by compressed air expansion are used to obtain the final CPCs. The axial and radial orientation during stretch blow molding improves the mechanical properties of CPCs in terms of impact strength, hardness, and rigidity, among others.

1.4 PROPERTIES OF CONDUCTIVE POLYMER COMPOSITES

Carbon nanofillers can be introduced into the insulating polymer matrix to endow the polymer with functionality. The prepared CPCs by this method have attracted increasing attention due to their improved mechanical properties, controllable electrical conductivities, excellent thermal properties, and enhanced barrier characteristics [37,38].

Introduction

1.4.1 MECHANICAL PROPERTIES

The addition of carbon nanofillers greatly affects the mechanical properties of CPCs. Compared to pure polymers, the elongation at break of polymers with carbon nanofillers often decreases, while the breaking strength, Young's modulus, and tensile strength all increase [2]. This may be attributed to the enhanced interfacial density of carbon nanofillers, as well as the interaction between carbon nanofillers and polymer matrices. The interface between carbon nanofillers and polymer matrix becomes well bonded, leading to effective stress transfer under strain [39].

1.4.2 ELECTRICAL PROPERTIES

Percolation theory can be used to describe the electrical properties of CPCs [23,37,40]. For polymers containing low contents of carbon nanofillers, the carbon nanofillers are isolated from each other within the polymer, making it difficult for electrons to pass through the carbon nanofillers, leading to high resistance composites. The conductivity of CPCs increases with the content of carbon nanofillers according to the classic percolation behavior. At a certain content of carbon nanofillers, the conductivity of CPCs enhances significantly to reach the percolation threshold of CPCs. The percolation threshold depends largely on aspects like the processing method of CPCs, the aspect ratio of carbon nanofillers, and the viscosity of the matrix. Further increase in the content of carbon nanofillers would progressively complete the conductive network structure constructed within CPCs, resulting in gradual slowing of conductivity of CPCs until reaching a constant. The preparation, processing, structure, and percolation threshold of carbon-based CPCs reported in recent years are listed in Table 1.1.

1.4.3 THERMAL PROPERTIES

The thermal properties of CPCs, including temperature effect, glass transition, viscous flow transition, melting transition, thermal stability, thermal expansion, and heat conduction, are greatly affected by the type and dispersion of nanofillers, as well as the interaction between nanofillers and polymer matrix. Guo et al. [49] constructed fully carbon nanofillers with a hierarchical structure suitable to fabricate highly thermally conductive polyimide nanocomposites. The obtained nanocomposites delivered a maximum conductivity λ of 1.60 W m^{-1} K^{-1} at a relatively low loading of f-MWCNT-g-RGO fillers (10 wt%, mass ratio of RGO to f-MWCNT of 2:1). However, some problems still require solutions, including the improvement of the trade-off between high mechanical properties and elevated thermal conductivity of CPCs. The presence of synergistic effects between nanofillers may also improve the thermal conductivity of CPCs. Additionally, the effect of surface treatment and modification on interfacial thermal resistance may also play certain roles.

1.4.4 BARRIER PROPERTIES

The barrier properties of polymer materials affect their practical applications in commercial production, such as anti-static packaging and storage of electronic devices.

TABLE 1.1
Preparation, Processing, Structure, and Percolation Threshold of Carbon-Based CPCs

Matrix	Nanofiller	Preparation	Processing	Structure	Percolation Threshold	References
PDMS	CNTs	Solution mixing	Casting	Self-segregation	0.003 vol%	[27]
TPU/SBS	CNTs	Melt mixing	Compression molding	Double percolated	0.38 wt%	[41]
XSBR	SSCNTs	Solution mixing	Compression molding	/	0.504 wt%	[42]
Bandage/PDMS	CNTs	Solution mixing	/	Self-segregation	0.51 wt%	[43]
SPX/TPU	CNTs	Solution mixing	/	Core-sheath	0.1 wt%	[44]
SPX/TPU	P-CNTs	Solution mixing	/	Core-sheath	0.1 wt%	[45]
PDMS	GO	Solution mixing	/	/	0.83 vol%	[14]
TPU	CNTs	Melt mixing	/	/	2 wt%	[46]
SEBS	GNPs	/	Compression molding	Segregation	0.299 vol%	[47]
SBR	CNTs	/	Compression molding	Segregation	0.035 vol%	[48]

CPCs, conductive polymer composites; GO, graphene oxide; PDMS, polydimethylsiloxane; SBR, boronic ester-cross-linked styrene−butadiene rubber; SEBS, styrene-b-(ethylene-co-butylene)-b-styrene triblock copolymer; SPX, spandex fiber; SSCNTs, hydrophilic sericin non-covalently modified carbon nanotubes; TPU, thermoplastic polyurethane; XSBR, carboxylic styrene−butadiene rubber.

Common barrier polymer materials include PE, PP, and polyvinyl alcohol. These materials suffer from disadvantages, such as low mechanical strengths and poor barriers to small molecules. Carbon nanofillers have large aspect ratios and specific surface areas, which can build stable conductive networks in the polymer matrices and form tortuous molecular movement paths. In addition, interfacial interactions exist between carbon nanofillers and polymer matrix, which can inhibit the migration of small molecules and improve the barrier performance of the composites.

1.5 SPECIALLY DESIGNED STRUCTURE

In addition to uniform dispersion structures of carbon nanofillers, some special structures constructed in CPCs have been designed, including the segregated structure and double percolated structure. These structures greatly impact the properties of the composites, especially their electrical properties. Segregated and double percolated structures can significantly reduce the percolation threshold of CPCs by regulating the distribution of percolation networks.

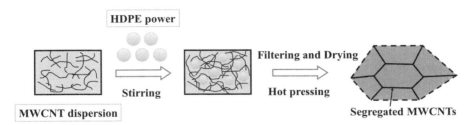

FIGURE 1.1 A schematic representation of the preparation process of CNT/HDPE composites with segregated structure. CNTs, carbon nanotubes; HDPE, high-density polyethylene.

1.5.1 SEGREGATED STRUCTURE

Segregated structures can be used to control the dispersion state of carbon nanofillers through the repulsion of polymer particles so that nanofillers may selectively be distributed between polymer particles. Since the nanofillers need distribution in only the slit between polymer particles to form conductive paths and not randomly distributed in the matrix, the percolation threshold of the CPCs may greatly be reduced. Hot pressing is a common method used to prepare CPCs with segregated structures. Xiang et al. [50] directly hot-pressed CNTs, GNPs, and HDPE powder at a high temperature. As shown in Figure 1.1, high viscosity HDPE was employed to distribute CNTs and GNPs in the gap of HDPE particles to form a segregated structure. The resulting special structure reduced the percolation thresholds of CPCs to 0.1 wt% CNTs and 2 wt% GNPs.

1.5.2 DOUBLE PERCOLATED STRUCTURE

In incompatible two-phase blend polymers, the carbon nanofillers are distributed in one phase through appropriate processing technology to yield a continuous phase in the whole composite system. The double percolated system includes two percolated behaviors: (i) percolation of carbon nanofillers in the conductive phase and (ii) percolation of the conductive phase in the whole composite. Xiang et al. [41] added CNTs to TPU blending with SBS. They then selected appropriate processing technology to control the selective distribution of CNTs in the composites to construct a double percolated structure (Figure 1.2). The conductivity test results showed a percolation threshold of CNT/TPU@SBS (CNTs distributed in TPU phase) of 0.38 wt%. This value was greatly reduced.

1.6 APPLICATIONS OF CONDUCTIVE POLYMER COMPOSITES IN FLEXIBLE STRAIN SENSORS

CPCs have widely been applied in flexible strain sensors due to their excellent stretchability, high mechanical flexibility, controllable electrical properties, and low preparation cost. In recent years, some CPC-based flexible strain sensors have

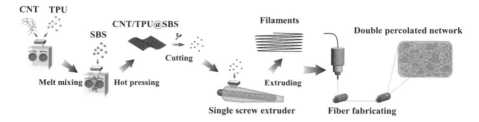

FIGURE 1.2 A schematic diagram of the preparation process and double percolated network of CNT/TPU@SBS. CNTs, carbon nanotubes; SBS, styrene butadiene styrene; TPU, thermoplastic polyurethane.

been fabricated. For instance, Wang et al. [51] prepared polydopamine-encapsulated CNTs/elastic bands (EBs) to strain sensor with a strain detection range of 920% and excellent durability (10,000 cycles of loading/unloading). However, the sensitivity of the sensor was relatively low (GF of only 5.06 at strain 0%–200%). Wang et al. [27] prepared a strain sensor based on CNTs and PDMS with high sensitivity (maximum GF of about 150) and low strain detection range (0%–30%). Therefore, flexible strain sensors require further optimization and improvement of performance, and more focus should be paid to preparing low-cost strain sensors. So far, numerous low-cost processing methods for flexible strain sensors have been reported. In this chapter, more focus was paid to coating, electrostatic self-assembly, 3D printing, and chemical vapor deposition, among others.

1.6.1 Fabrication of Flexible Strain Sensors

1.6.1.1 Coating

The coating technology can be utilized to uniformly coat CPCs on the matrix material surface through a specific process. The coating technology is a simple process used for preparing flexible strain sensors [10,32,52]. Zhang et al. [53] coated CNT/TPU composites on commercially available spandex multifilament yarn to yield flexible strain sensors with a low percolation threshold (0.015 wt%). A CNT/PDMS composite coating with a self-segregated structure was prepared by Liu et al. [43]. In this case, the coating was coated on the elastic medical bandage to form a stretchable strain sensor with a low percolation threshold (0.51 wt%), wide strain detection range (0%–200%), and high sensitivity (GF of 615). Unlike the dispersion of nanofillers within the whole matrix, the coating formed conductive networks on the matrix surface, which can reduce its percolation threshold [44]. Moreover, the CPCs on the matrix surface can produce micro folds or micro-cracks by pre-stretching, conducive to regulating the performance of the strain sensors [54].

1.6.1.2 Electrostatic Self-Assembly

The materials used in electrostatic self-assembly are usually anions, cations, or polymers, and monomers with charged groups, charged metal nanoparticles, or inorganic nanoparticles. The nanofillers are often charged to provide heterogeneous charges. Based on the principle of heterosexual attraction, strain sensors can be formed by

Introduction

electrostatic self-assembly. Wu et al. [24] charged CB followed by adsorption on PU sponge by electrostatic adsorption to form a strain sensor. The obtained sensor was found suitable for subtle strain monitoring (91 Pa pressure, 0.2% strain) based on the microcrack junction sensing mechanism and compressive contact of CB@PU conductive backbones for large strain monitoring (16.4 kPa pressure, 60% strain).

1.6.1.3 3D Printing

3D printing is a technology based on digital model files using adhesive materials, such as metal powder or plastic to print objects layer by layer. Compared to the traditional processing method of CPC-based flexible strain sensors, 3D printing is advantageous in terms of high resolution, elevated accuracy, low cost, fast fabrication, and customization for mass production without the need for molds [55,56].

Currently, 3D printing technologies based on material extrusion, photocuring, and laser sintering involving fused filament fabrication (FFF), direct ink writing (DIW), digital light processing (DLP), and selective laser sintering (SLS) are widely used to fabricate CPC-based flexible strain sensors. For strain sensors prepared by FFF and DIW, the shear force generated during the processing decides the extruded composites' micro-orientation. Also, the nanofillers are well dispersed within the polymer matrix, resulting in flexible strain sensors with enhanced performance [57]. DLP is famous for its high efficiency, elevated resolution, and superior precision in the fabrication of flexible strain sensors. For example, Guo et al. [58] introduced carboxyl CNTs into the ACMO resin to fully mitigate the over-curing of ACMO resin and achieve a large strain detection range (up to 100%). However, some problems dealing with the preparation of flexible strain sensors using 3D printing still require solutions. These include pre-processing before printing, nozzle blockage, post-processing [20,59], materials used for printing limited to light-curing resin [60,61], powder pollution during printing, rough surface of printed products, and deformation of products stored for long periods due to internal stress [62]. Reducing the resin viscosity and UV absorption of nanofillers for DLP-based flexible strain sensors should also be considered.

1.6.1.4 Chemical Vapor Deposition

Chemical vapor deposition can be used to grow nanomaterials on the matrix surface through chemical evaporation under a vacuum. Chemical vapor deposition is advantageous in terms of simple deposition, convenient design of deposited nanomaterials, and deposition on the surfaces of complex structures. In this view, Hu et al. [63] prepared carbon hybrid fiber strain sensors by controllable wet spinning and appropriate chemical vapor deposition process, composed of a graphene fiber skeleton and CNT branches. The designed structure yielded fiber sensors with satisfactory performance in terms of high GF (up to 1,127), fast response time (<70 ms), and excellent reliability and stability (>2,000 cycles). However, chemical vapor deposition still suffers from slow deposition rates, toxic and harmful gases issued during the process, and difficulty in effectively modifying matrices locally.

1.6.1.5 Other Methods

Other common routes have also been employed to prepare CPC-based flexible strain sensors, including electrospinning, wet spinning, melt spinning, dip coating, and swelling. For example, Wang et al. [51] embedded CNTs into the EB by swelling-ultrasonication treatment to prepare strain sensors with high sensing performances. The resulting flexible strain sensors integrated a wide strain detection range (920% strain), high sensitivity (GF of 129 at 780% strain), excellent stability (10,000 cycles at 100% strain), and fast response. However, these methods are often limited by the long preparation cycle, unstable preparation process, and high-cost equipment, not conducive to large-scale industrial production. Therefore, developing simple, fast, and low-cost preparation methods for flexible strain sensors is highly desirable.

1.6.2 SENSING MECHANISM OF FLEXIBLE STRAIN SENSORS

The strain resistance response of traditional strain sensors is caused by the changes in material geometry under external force and intrinsic piezoresistive effects. Unlike traditional strain mechanisms, the strain sensing mechanism of CPC-based flexible strain sensors is affected by the type of materials, micro- and nano-structures, and preparation processes [11,21,49]. A total of five types of sensing mechanisms have been identified so far: (i) tunneling effect, (ii) geometric effect, (iii) piezoresistive effect, (iv) crack propagation, (v) and disconnection mechanism.

1.6.2.1 Tunneling Effect

For resistive-type CPC-based flexible strain sensors, low contents of carbon nanofillers within CPCs often lead to the formation of a polymer segregated layer between the nanofillers, playing a certain barrier in hindering the transport of electrons. The phenomenon by which electrons overcome the barrier to pass through the segregated layer is called the tunneling effect. The tunneling resistance between adjacent nanofillers can be approximately estimated by Simmons' tunneling resistance theory [64]. This may be expressed by Eq. (1.1):

$$R_{\text{tunneling}} = \frac{V}{AJ} = \frac{h^2 d}{Ae^2 \sqrt{2m\lambda}} \exp\left(\frac{4\pi d}{h}\sqrt{2m\lambda}\right) \quad (1.1)$$

where J represents the tunneling current density, A is the cross-sectional area of the nanofiller, V is the potential, m refers to the mass of electrons, e is the quantum of electricity, h denotes the Planck constant, d is the distance between the nanofiller and elastomer, and λ represents the energy barrier height of the elastomer.

The tunnel effect is the main mechanism involved in strain sensors with low content nanofillers under small strains. Hu et al. [65] comprehensively studied the tunneling effect of CNT/polymer strain sensors and noticed weak non-linear piezoresistance both experimentally and numerically when small CNT volume fractions were used.

1.6.2.2 Geometric Effect

For capacitive-type strain sensors, the geometric effect is the main involved sensing mechanism. When a stretchable material is stretched, it tends to stretch in the tensile direction and contract in the transverse direction based on Poisson's ratio of v. For conductors, the resistance can be expressed by Eq. (1.2):

Introduction

$$R = \frac{\rho L}{A} \tag{1.2}$$

where ρ represents the electrical resistivity, L is the length, and A is the cross-sectional area of the conductor.

When a conductor is stretched, its length increases while its cross-sectional area decreases, resulting in enhanced resistance. For capacitive-type sensors, the capacitance can be raised by incrementing the capacitance area and reducing the thickness of the dielectric layer. For a parallel plate capacitor with an initial length of l_0, a width of w_0, and a dielectric layer thickness of d_0, the initial capacitance value can be expressed by Eq. (1.3) [64]:

$$C_0 = \varepsilon_0 \varepsilon_r \frac{l_0 w_0}{d_0} \tag{1.3}$$

where ε_0 and ε_r are the electric and dielectric constant of the dielectric layer, respectively.

At strain ε, the length l is stretched to $(1+\varepsilon)l_0$, while width and thickness are contracted to $(1-v_{electrode}\varepsilon)w_0$ and $(1-v_{dielectric}\varepsilon)d_0$, respectively. Here, $v_{elctrode}$ and $v_{dielectric}$ represent the Poisson's ratios of the electrode and dielectric layer under tensile, respectively. For retractable sensors, the Poisson's ratio of the retractable electrodes and the dielectric layers are assumed to be the same. The capacitance under tensile can be expressed by Eq. (1.4) [64]:

$$C_0 = \varepsilon_0 \varepsilon_r \frac{(1+\varepsilon)l_0 \,(1-v_{electrode}\varepsilon)w_0}{(1-v_{dielectric}\varepsilon)d_0} \tag{1.4}$$

According to Eq. (1.4), the capacitance of capacitive-type sensors should be a linear function. However, if the strain is too large, the linear relationship between different axes through the Poisson's ratio is invalid for the polymer. As a result, the capacitance and strain become linear within a certain range of strain.

1.6.2.3 Piezoresistive Effect

The piezoresistive effect refers to the change in material resistivity due to the variation in the conductive network structure within the material under strain. The main function of the strain sensor is feeding back the applied strain by changing sensor resistivity. CPC-based flexible strain sensors can be divided into three piezoresistive characteristics. The first piezoresistive characteristic states no change in resistance rate of the sensor after compression deformation since the addition of less carbon nanofiller to the conductive composite is not enough for the sensor to form an effective conductive path after compression. Therefore, the sensor remains always at a high resistance level before and after the compression strain. The second piezoresistive characteristic consists of larger contents of carbon nanofillers, where the dense distribution of nanofillers within the polymer matrix can change the original conductive network structure and increase the conductive path under small compressive strains. Therefore, the resistance of the strain sensor under external force decreases rapidly to remain at a low resistance level. The third piezoresistive characteristic has to do with the resistance of the strain sensor, which decreases with the increase in external force to show an obvious changing process. The reason for this has to do with the moderate content of carbon nanofillers so that the strain sensor has a wide sensing response range. Therefore, the addition of nanofillers during the preparation of CPC flexible strain sensors should yield strain sensors with high conductivity and excellent mechanical properties, coupled with good piezoresistive characteristics and a large strain detection range.

1.6.2.4 Crack Propagation
Crack propagation often affects the performance of flexible strain sensors since processing causes internal defects or stress concentration of CPCs, with cracks occurring at the defects or stress concentration under strain application. The crack often enhances with the strain until CPCs fracture. Meanwhile, the conductivity of CPCs declines with the crack propagation and recovers with the crack closure [54]. The crack propagation mechanism is mostly used in the coating of polymer matrix surfaces to design flexible strain sensors with high sensitivity.

1.6.2.5 Disconnection Mechanism
In CPCs, carbon nanofillers construct conductive networks, and electrons can pass through the conductive networks to provide CPCs with good conductivity. Under strain, the movement of the polymer matrix affects the internal carbon nanofillers, and the conductive networks composed of nanofillers are gradually destroyed, resulting in the progressively increased resistance of CPCs [66]. For microstructure, the reason for the disconnection in conductive networks under strain has to do with the slip of nanofillers due to the weak interface adhesion and stiffness mismatch between rigid nanofillers and polymer matrix [45].

1.6.3 SENSING PERFORMANCES

The performance parameters of flexible strain sensors include but are not limited to strain detection range, sensitivity (such as GF), linearity, response time, hysteresis, dynamic sensing performances, and durability. Nanofillers could be used to adjust the microstructure, thereby optimizing the performance parameters of flexible strain sensors.

1.6.3.1 Strain Detection Range
Strain detection range is a key sensing performance parameter of flexible strain sensors, which determines the actual working range of flexible strain sensors. Strain detection range is related to the polymer matrix materials, nanofillers, and the structure of flexible strain sensors. Polymer matrix materials with high stretchability usually include PU (TPU), PDMS, SBS, natural rubber (NR) [67], PVDF [68], polyethylene terephthalate (PET) [69], polyimide (PI) [70], and Ecoflex [71]. The Young's modulus of PDMS is much lower (around 0.4–3.5 MPa) when compared to other matrices, such as PU (15.1–151.4 MPa for 0.16 and 0.32 g cm^{-3} foam, respectively), PI (84.1 GPa with 3.25% elongation at break), and PVDF (0.84 GPa

Introduction

with 0.86% elongation at break) [64]. The low Young's modulus of PDMS suggests its elastic matrix has better flexibility and versatility than other polymers. Moreover, the high Poisson's ratio of one-dimensional nanofillers facilitates the formation of effective percolation networks. Even at high strain levels, the flexible strain sensors could maintain stable electromechanical characteristics. Specific structures or processing may also increase the stretchability of the strain sensors. Examples include fold structures, helical structures, and encapsulation, among others [6,19,31].

1.6.3.2 Sensitivity

The relationship between the relative changes in electrical signal and applied strain reflects the sensitivity of strain sensors. Sensitivity can be expressed according to Eq. (1.5) [16,61]:

$$\mathrm{GF} = \frac{\Delta R}{R_0 \varepsilon} \left(\mathrm{GF} = \frac{\Delta I}{I_0 \varepsilon} \right),$$

$$= \frac{\Delta C}{C_0 \varepsilon} \tag{1.5}$$

where ΔR, ΔI, and ΔC reflect the change in resistance, current, and capacitance. R_0, I_0, and C_0 are the initial resistance, current, and capacitance at strain $\varepsilon = 0\%$. ε is the applied strain.

The material type, agglomeration, dispersion of nanofillers, macroscopic and microscopic structures of strain sensors, and different processing technologies all determine the percolated network structures of the strain sensors in terms of sensitivity and other sensor characteristics. High-density percolation networks lead to nanofillers in contact with each other, resulting in dominant contact resistance and a robust percolation network. The decrease in percolation network density separates more nanofillers from each other, and contact resistance is gradually transformed into tunneling resistance. This facilitates the contact nodes between nanofillers to become disconnected, resulting in higher sensitivity composites. When the percolation networks are at low density, most nanofillers are subject to the state of tunneling effect. The tunneling resistance becomes dominant, leading to a fragile percolation network, which makes the sensitivity of composite increases sharply. Overall, a robust percolated network structure is needed to achieve a wide strain detection range, while a fragile percolated network structure is more conducive to achieving high sensitivity [65]. Thus, the trade-off between a wide strain detection range and high sensitivity is still challenging.

1.6.3.3 Linearity

Linearity affects the calibration of flexible strain sensors. High linearity in the strain detection range may prevent complex calibration and ensure accurate detection of complex strain states [31]. When the microstructure of a flexible strain sensor is gradually oriented from "uniform morphology", the strain sensor often produces a non-linear response as a function of the applied strain. The transformation of the conductive network from a uniform state to a non-uniform state is usually the reason behind the non-linear strain responses of sensors.

1.6.3.4 Hysteresis

Hysteresis also impacts the practical applications of sensors. Flexible strain sensors can be used as wearable electronic devices to monitor dynamic motions, such as walking, running, speaking, heartbeat, and breathing [49,72,73]. For capacitive-type strain sensors, the hysteresis is usually negligible. For resistive-type strain sensors, the stretchable conductive network requires time to slide back to its original state, leading to large hysteresis under strain. The value of hysteresis (H_m) can be expressed by Eq. (1.6) [51]:

$$H_m = \frac{A_s - A_R}{A_S} \tag{1.6}$$

where A_S and A_R are the areas under the stretching and releasing curves in the stress-strain curves, respectively.

The hysteresis effect can be strengthened by increasing strain. Hysteresis is mainly caused by the viscoelasticity of polymers, as well as the interaction between nanofillers and polymer matrix. For weak interaction between nanofillers and matrix, nanofillers could slide into the polymer matrix under large strain, but they cannot slide into the original position quickly after fully releasing the strain, resulting in high hysteresis behavior. Strong interactions between nanofillers and polymer matrices exist, leading to a low hysteresis effect of strain sensors prepared by nanocomposites.

1.6.3.5 Dynamic Durability

Dynamic durability relies on the stable electromechanical performance of strain sensors during long-term loading/unloading cycles. Dynamic durability is important for flexible strain sensors when used as wearable electronic devices since excellent dynamic durability would facilitate adaption to various complex environments [29,66]. The degradation of strain sensor performance is mainly caused by the fatigue and plastic deformation of the polymer matrix under high strain, as well as the fracture and deformation of nanomaterials.

1.7 CONCLUSIONS

In recent years, carbon-based CPCs have attracted increasing interest due to their simple preparation methods, high flexibility, low cost, and excellent properties. Carbon-based CPCs have great potential applications in anti-static protection, electrostatic dissipation, electromagnetic interference shielding, and flexible strain sensors. The properties of carbon-based CPCs depend on the distribution and dispersion of carbon nanofillers in the polymer matrix, related to factors like the preparation methods, processing, and structures of carbon-based CPCs. In this chapter, the preparation methods, processing, performances, and structures of carbon-based CPCs, as well as the applications of carbon-based CPCs in flexible strain sensors were discussed. Particular attention was paid to the preparation methods, sensing mechanisms, and sensing performances of CPC-based flexible strain sensors.

REFERENCES

1. Wang T, Zhang Y, Liu Q, Cheng W, Wang X, Pan L, et al. A self-healable, highly stretchable, and solution processable conductive polymer composite for ultrasensitive strain and pressure sensing. *Adv Funct Mater.* 2018;28(7):1705551.
2. Pan L, Wang F, Cheng Y, Leow WR, Zhang YW, Wang M, et al. A supertough electrotendon based on spider silk composites. *Nat Commun.* 2020;11(1):1332.
3. Sun Z, Yang S, Zhao P, Zhang J, Yang Y, Ye X, et al. Skin-like ultrasensitive strain sensor for full-range detection of human health monitoring. *ACS Appl Mater Inter.* 2020;12(11):13287–95.
4. Yue X, Jia Y, Wang X, Zhou K, Zhai W, Zheng G, et al. Highly stretchable and durable fiber-shaped strain sensor with porous core-sheath structure for human motion monitoring. *Compos Sci Technol.* 2020;189:108038.
5. Li J, Wang L, Wang X, Yang Y, Hu Z, Liu L, et al. Highly conductive PVA/Ag coating by aqueous in situ reduction and its stretchable structure for strain sensor. *ACS Appl Mater Inter.* 2020;12(1):1427–35.
6. Yang Z, Zhai Z, Song Z, Wu Y, Liang J, Shan Y, et al. Conductive and elastic 3D helical fibers for use in washable and wearable electronics. *Adv Mater.* 2020;32(10):e1907495.
7. Zhang R, Lv A, Ying C, Hu Z, Hu H, Chen H, et al. Facile one-step preparation of laminated PDMS based flexible strain sensors with high conductivity and sensitivity via filler sedimentation. *Compos Sci Technol.* 2020;186:107933.
8. Lan L, Jiang C, Yao Y, Ping J, Ying Y. A stretchable and conductive fiber for multifunctional sensing and energy harvesting. *Nano Energy.* 2021;84:105954.
9. Zheng Y, Li Y, Li Z, Wang Y, Dai K, Zheng G, et al. The effect of filler dimensionality on the electromechanical performance of polydimethylsiloxane based conductive nanocomposites for flexible strain sensors. *Compos Sci Technol.* 2017;139:64–73.
10. Li Y, Zhou B, Zheng G, Liu X, Li T, Yan C, et al. Continuously prepared highly conductive and stretchable SWNT/MWNT synergistically composited electrospun thermoplastic polyurethane yarns for wearable sensing. *J Mater Chem C.* 2018;6(9):2258–69.
11. Chen J, Zhu Y, Jiang W. A stretchable and transparent strain sensor based on sandwich-like PDMS/CNTs/PDMS composite containing an ultrathin conductive CNT layer. *Compos Sci Technol.* 2020;186:107938.
12. He Y, Wu D, Zhou M, Zheng Y, Wang T, Lu C, et al. Wearable strain sensors based on a porous polydimethylsiloxane hybrid with carbon nanotubes and graphene. *ACS Appl Mater Inter.* 2021;13(13):15572–83.
13. Zhang D, Chi B, Li B, Gao Z, Du Y, Guo J, et al. Fabrication of highly conductive graphene flexible circuits by 3D printing. *Synth Met.* 2016;217:79–86.
14. Shi G, Lowe SE, Teo AJT, Dinh TK, Tan SH, Qin J, et al. A versatile PDMS submicrobead/graphene oxide nanocomposite ink for the direct ink writing of wearable micronscale tactile sensors. *Appl Mater Today.* 2019;16:482–92.
15. Zhang H, Liu D, Lee JH, Chen H, Kim E, Shen X, et al. Anisotropic, wrinkled, and crack-bridging structure for ultrasensitive, highly selective multidirectional strain sensors. *Nanomicro Lett.* 2021;13(1):122.
16. Ma J, Wang P, Chen H, Bao S, Chen W, Lu H. Highly sensitive and large-range strain sensor with a self-compensated two-order structure for human motion detection. *ACS Appl Mater Inter.* 2019;11(8):8527–36.
17. Xiang D, Harkin-Jones E, Linton D, Martin P. Structure, mechanical, and electrical properties of high-density polyethylene/multi-walled carbon nanotube composites processed by compression molding and blown film extrusion. *J Appl Polym Sci.* 2015;132(42):42665.
18. Yuan Q, Wu D. Low percolation threshold and high conductivity in carbon black filled polyethylene and polypropylene composites. *J Appl Polym Sci.* 2010;115(6):3527–34.

19. Wang L, Chen Y, Lin L, Wang H, Huang X, Xue H, et al. Highly stretchable, anticorrosive and wearable strain sensors based on the PDMS/CNTs decorated elastomer nanofiber composite. *Chem Eng J.* 2019;362:89–98.
20. Christ JF, Aliheidari N, Potschke P, Ameli A. Bidirectional and stretchable piezoresistive sensors enabled by multimaterial 3D printing of carbon nanotube/thermoplastic polyurethane nanocomposites. *Polymers (Basel).* 2018;11(1):11–26.
21. Yu S, Wang X, Xiang H, Zhu L, Tebyetekerwa M, Zhu M. Superior piezoresistive strain sensing behaviors of carbon nanotubes in one-dimensional polymer fiber structure. *Carbon.* 2018;140:1–9.
22. Yu S, Wang X, Xiang H, Tebyetekerwa M, Zhu M. 1-D polymer ternary composites: Understanding materials interaction, percolation behaviors and mechanism toward ultra-high stretchable and super-sensitive strain sensors. *Sci China Mater.* 2019;62(7):995–1004.
23. Ji M, Deng H, Yan D, Li X, Duan L, Fu Q. Selective localization of multi-walled carbon nanotubes in thermoplastic elastomer blends: An effective method for tunable resistivity–strain sensing behavior. *Compos Sci Technol.* 2014;92:16–26.
24. Wu X, Han Y, Zhang X, Zhou Z, Lu C. Large-area compliant, low-cost, and versatile pressure-sensing platform based on microcrack-designed carbon black@polyurethane sponge for human-machine interfacing. *Adv Funct Mater.* 2016;26(34):6246–56.
25. Wang Y, Hao J, Huang Z, Zheng G, Dai K, Liu C, et al. Flexible electrically resistive-type strain sensors based on reduced graphene oxide-decorated electrospun polymer fibrous mats for human motion monitoring. *Carbon.* 2018;126:360–71.
26. Ravindren R, Mondal S, Nath K, Das NC. Synergistic effect of double percolated co-supportive MWCNT-CB conductive network for high-performance EMI shielding application. *Polym Adv Technol.* 2019;30(6):1506–17.
27. Wang M, Zhang K, Dai XX, Li Y, Guo J, Liu H, et al. Enhanced electrical conductivity and piezoresistive sensing in multi-wall carbon nanotubes/polydimethylsiloxane nanocomposites via the construction of a self-segregated structure. *Nanoscale.* 2017;9(31):11017–26.
28. Xiang D, Wang L, Tang Y, Zhao C, Harkin-Jones E, Li Y. Effect of phase transitions on the electrical properties of polymer/carbon nanotube and polymer/graphene nanoplatelet composites with different conductive network structures. *Polym Int.* 2018;67(2):227–35.
29. Li X, Fan YJ, Li HY, Cao JW, Xiao YC, Wang Y, et al. Ultracomfortable hierarchical nanonetwork for highly sensitive pressure sensor. *ACS Nano.* 2020;14(8):9605–12.
30. Zhuang Y, Guo Y, Li J, Jiang K, Yu Y, Zhang H, et al. Preparation and laser sintering of a thermoplastic polyurethane carbon nanotube composite-based pressure sensor. *RSC Adv.* 2020;10(40):23644–52.
31. Zhang H, Liu D, Lee J-H, Chen H, Kim E, Shen X, et al. Anisotropic, wrinkled, and crack-bridging structure for ultrasensitive, highly selective multidirectional strain sensors. *Nano-Micro Lett.* 2021;13(1):122.
32. Zhang M, Wang C, Wang Q, Jian M, Zhang Y. Sheath-core graphite/silk fiber made by dry-meyer-rod-coating for wearable strain sensors. *ACS Appl Mater Inter.* 2016;8(32):20894–9.
33. Wang X, Wan A, Jiang G, Raji RK, Yu D. Preparation of polypyrrole/silver conductive polyester fabric by UV exposure. *Autex Res J.* 2020;21(2):231–237.
34. Su C, Xu L, Zhang C, Zhu J. Selective location and conductive network formation of multiwalled carbon nanotubes in polycarbonate/poly(vinylidene fluoride) blends. *Compos Sci Technol.* 2011;71(7):1016–21.
35. Mayoral B, Menary G, Martin P, Garrett G, Millar B, Douglas P, et al. Characterizing biaxially stretched polypropylene/graphene nanoplatelet composites. *Front Mater.* 2021;8:687282.

36. Vadukumpully S, Paul J, Mahanta N, Valiyaveettil S. Flexible conductive graphene/poly(vinyl chloride) composite thin films with high mechanical strength and thermal stability. *Carbon.* 2011;49(1):198–205.
37. Zhang S, Deng H, Zhang Q, Fu Q. Formation of conductive networks with both segregated and double-percolated characteristic in conductive polymer composites with balanced properties. *ACS Appl Mater Inter.* 2014;6(9):6835–44.
38. Brook I, Tchoudakov R, Suckeveriene RY, Narkis M. Electro-mechanical sensors based on conductive hybrid nanocomposites. *Polym Adv Technol.* 2015;26(7):889–97.
39. Arif MF, Alhashmi H, Varadarajan KM, Koo JH, Hart AJ, Kumar S. Multifunctional performance of carbon nanotubes and graphene nanoplatelets reinforced PEEK composites enabled via FFF additive manufacturing. *Compos B: Eng.* 2020;184:107625.
40. Gao X, Zhang S, Mai F, Lin L, Deng Y, Deng H, et al. Preparation of high performance conductive polymer fibres from double percolated structure. *J Mater Chem.* 2011;21(17):6401–6408.
41. Xiang D, Liu L, Chen X, Wu Y, Wang M, Zhang J, et al. High-performance fiber strain sensor of carbon nanotube/thermoplastic polyurethane@styrene butadiene styrene with a double percolated structure. *Front Mater Sci.* 2022;16(1):220586.
42. Lin M, Zheng Z, Yang L, Luo M, Fu L, Lin B, et al. A high-performance, sensitive, wearable multifunctional sensor based on rubber/CNT for human motion and skin temperature detection. *Adv Mater.* 2021;34(1):e2107309.
43. Liu L, Zhang X, Xiang D, Wu Y, Sun D, Shen J, et al. Highly stretchable, sensitive and wide linear responsive fabric-based strain sensors with a self-segregated carbon nanotube (CNT)/polydimethylsiloxane (PDMS) coating. *Prog Nat Sci.* 2022;32:34–42.
44. Chen Q, Xiang D, Wang L, Tang Y, Harkin-Jones E, Zhao C, et al. Facile fabrication and performance of robust polymer/carbon nanotube coated spandex fibers for strain sensing. *Compos Part A Appl Sci Manuf.* 2018;112:186–96.
45. Chen Q, Li Y, Xiang D, Zheng Y, Zhu W, Zhao C, et al. Enhanced strain sensing performance of polymer/carbon nanotube-coated spandex fibers via noncovalent interactions. *Macromol Mater Eng.* 2019;305(2):1900525.
46. Christ JF, Aliheidari N, Ameli A, Pötschke P. 3D printed highly elastic strain sensors of multiwalled carbon nanotube/thermoplastic polyurethane nanocomposites. *Mater Design.* 2017;131:394–401.
47. Shen Z, Feng J. Mass-produced SEBS/graphite nanoplatelet composites with a segregated structure for highly stretchable and recyclable strain sensors. *J Mater Chem C.* 2019;7(30):9423–9.
48. Huang Q, Tang Z, Wang D, Wu S, Guo B. Engineering segregated structures in a cross-linked elastomeric network enabled by dynamic cross-link reshuffling. *ACS Macro Lett.* 2021;10(2):231–6.
49. Guo Y, Ruan K, Yang X, Ma T, Kong J, Wu N, et al. Constructing fully carbon-based fillers with a hierarchical structure to fabricate highly thermally conductive polyimide nanocomposites. *J Mater Chem C.* 2019;7(23):7035–44.
50. Xiang D, Wang L, Tang Y, Harkin-Jones E, Zhao C, Wang P, et al. Damage self-sensing behavior of carbon nanofiller reinforced polymer composites with different conductive network structures. *Polym.* 2018;158:308–19.
51. Wang Y, Jia Y, Zhou Y, Wang Y, Zheng G, Dai K, et al. Ultra-stretchable, sensitive and durable strain sensors based on polydopamine encapsulated carbon nanotubes/elastic bands. *J Mater Chem C.* 2018;6(30):8160–70.
52. Qi K, Zhou Y, Ou K, Dai Y, You X, Wang H, et al. Weavable and stretchable piezoresistive carbon nanotubes-embedded nanofiber sensing yarns for highly sensitive and multimodal wearable textile sensor. *Carbon.* 2020;170:464–76.
53. Zhang R, Deng H, Valenca R, Jin J, Fu Q, Bilotti E, et al. Carbon nanotube polymer coatings for textile yarns with good strain sensing capability. *Sens Actuator A Phys.* 2012;179:83–91.

54. Wan K, Liu Y, Santagiuliana G, Barandun G, Taroni Junior P, Güder F, et al. Self-powered ultrasensitive and highly stretchable temperature–strain sensing composite yarns. *Mater Horizons*. 2021;8(9):2513–2519.
55. Zhang J, Ye S, Liu H, Chen X, Chen X, Li B, et al. 3D printed piezoelectric BNNTs nanocomposites with tunable interface and microarchitectures for self-powered conformal sensors. *Nano Energy*. 2020;77:105300.
56. Liu H, Zhang H, Han W, Lin H, Li R, Zhu J, et al. 3D printed flexible strain sensors: From printing to devices and signals. *Adv Mater*. 2021;33(8):e2004782.
57. Xiang D, Zhang Z, Han Z, Zhang X, Zhou Z, Zhang J, et al. Effects of non-covalent interactions on the properties of 3D printed flexible piezoresistive strain sensors of conductive polymer composites. *Compos Inter*. 2021;28(6):577–91.
58. Guo B, Ji X, Chen X, Li G, Lu Y, Bai J. A highly stretchable and intrinsically self-healing strain sensor produced by 3D printing. *Virtual and Physical Prototyping*. 2020;15(supl):520–31.
59. Abshirini M, Charara M, Marashizadeh P, Saha MC, Altan MC, Liu Y. Functional nanocomposites for 3D printing of stretchable and wearable sensors. *Appl Nanosci*. 2019;9(8):2071–83.
60. Peng S, Blanloeuil P, Wu S, Wang CH. Rational design of ultrasensitive pressure sensors by tailoring microscopic features. *Adv Mater Inter*. 2018;5(18):1800403.
61. Yin XY, Zhang Y, Xiao J, Moorlag C, Yang J. Monolithic dual-material 3D printing of ionic skins with long-term performance stability. *Adv Funct Mater*. 2019;29(39):1904716.
62. Liu L, Xiang D, Wu Y, Zhou Z, Li H, Zhao C, et al. Conductive polymer composites based flexible strain sensors by 3D printing: A mini-review. *Front Mater*. 2021;8:725420.
63. Hu Y, Huang T, Zhang H, Lin H, Zhang Y, Ke L, et al. Ultrasensitive and wearable carbon hybrid fiber devices as robust intelligent sensors. *ACS Appl Mater Inter*. 2021;13(20):23905–14.
64. Lu Y, Biswas MC, Guo Z, Jeon JW, Wujcik EK. Recent developments in bio-monitoring via advanced polymer nanocomposite-based wearable strain sensors. *Biosens Bioelectron*. 2019;123:167–77.
65. Hu B, Hu N, Li Y, Akagi K, Yuan W, Watanabe T, et al. Multi-scale numerical simulations on piezoresistivity of CNT/polymer nanocomposites. *Nanoscale Res Lett*. 2012;7(1):402.
66. He Y, Wu D, Zhou M, Zheng Y, Wang T, Lu C, et al. Wearable strain sensors based on a porous polydimethylsiloxane hybrid with carbon nanotubes and graphene. *ACS Appl Mater Inter*. 2021;13(13):15572–15583.
67. Huang P, Xia Z, Cui S. 3D printing of carbon fiber-filled conductive silicon rubber. *Mater Design*. 2018;142:11–21.
68. Zhao M, Ren Z-Z, Yang M-B, Yang W. Effects of modified nano-silica on the microstructure of PVDF and its microporous membranes. *J Polym Res*. 2019;26(2):1.
69. Fang Y, Liu X, Wang C. Layer-by-layer assembly flame-retardant and anti-dripping treatment of polyethylene terephthalate fabrics. *J Eng Fiber Fabr*. 2019;14:1–8.
70. Qiu G, Ma W, Wu L. Low dielectric constant polyimide mixtures fabricated by polyimide matrix and polyimide microsphere fillers. *Polym Int*. 2020;69(5):485–91.
71. Liao Z, Hossain M, Yao X. Ecoflex polymer of different Shore hardnesses: Experimental investigations and constitutive modelling. *Mech Mater*. 2020;144:103366.
72. Zhang X, Xiang D, Zhu W, Zheng Y, Harkin-Jones E, Wang P, et al. Flexible and high-performance piezoresistive strain sensors based on carbon nanoparticles@polyurethane sponges. *Compos Sci Technol*. 2020;200:108437.
73. Cao Y, Lai T, Teng F, Liu C, Li A. Highly stretchable and sensitive strain sensor based on silver nanowires/carbon nanotubes on hair band for human motion detection. *Prog Nat Sci*. 2021;31(3):379–86.

2 Compression Molded Conductive Polymer Composites

2.1 INTRODUCTION

Compression molding process is a process method that puts the material into a mold cavity of the metal mold and uses a certain temperature and pressure generated by the hot press to soften the material in the mold cavity, flow under pressure, fill the mold cavity and solidify to obtain composite components [1] (Figure 2.1).

Composite components need to go through four necessary processes from design to manufacture. The first is the structural design of composite components, that is, the structural design is carried out according to the service requirements of components; the second is the process design of composite components. The structural characteristics, process feasibility, molding mode, and clamping sequence of components should be fully considered in the design of forming die, and then the manufacturing and acceptance of the forming mold of composite components. The mold shall be processed according to the design requirements and can be used only after being manufactured and passing the inspection. The last is the molding of composite components. Composite products can be obtained after feeding, mold closing, curing/melting, demolding, trimming, and passing the inspection.

Compression molding has good observability, high dimensional accuracy, and easy to ensure internal quality, so it is widely used in the manufacture of composite components with complex profile. Therefore, compression molding plays an important role in composite molding process.

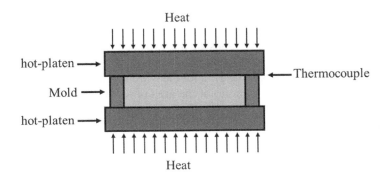

FIGURE 2.1 Schematic of compression molding the mold to obtain the compression molded component.

DOI: 10.1201/9781003218661-2

2.2 COMPRESSION MOLDED HDPE/MWCNT COMPOSITES

When molding plastic components, the influence of processing conditions on the structure and properties of molded polymers is very important for industrial applications. If the structure is affected by processing, its performance will also be affected. Therefore, it is particularly important to study the processing conditions of molding process and the structure, mechanical properties, and electrical properties of polymer/carbon nanofiller composites.

High-density polyethylene (HDPE) is an important low-cost commodity thermoplastic. Carbon nanotubes (CNTs) are widely used because of their large aspect ratio and excellent mechanical properties. If the performance of HDPE can be improved by adding CNTs, the application range of this material may be greatly expanded, such as in the production of conductive packaging film. However, the strong van der Waals force between CNTs limits their separation and dispersion in the polymer matrix. The existence of a large number of CNT aggregates will lead to the decline in mechanical and electrical properties of materials. In the past few years, the dispersion, structure, and properties of melt-mixed HDPE/CNT nanocomposites have been widely discussed. The uniform dispersion of CNTs is an important prerequisite for the successful preparation of polymer/CNT nanocomposites with ideal properties.

2.2.1 Polarized Optical Microscopy

In order to investigate the macrodispersion of primary nanotube agglomerates in the HDPE after melt mixing, polarized optical microscopy (POM) was performed for both extruded and compression molded samples. It was not possible to conduct POM tests for the nanocomposites with 8 wt% multi-walled carbon nanotubes (MWCNTs) as light could not penetrate the sample due to the high density of MWCNTs dispersed in the matrix. The microscopic images of the extruded and compression molded composites with 1, 2, and 4 wt% MWCNTs are shown in Figure 2.2a–c and d–f, respectively. The POM images show that the primary MWCNT agglomerates are distributed homogeneously in the HDPE matrix and more primary agglomerates with increasing MWCNT content can be observed in both extruded and compression molded samples. It can be seen that the compression molding process has no obvious effect on the state of primary agglomerates by comparing the micrographs of the extruded and compression molded composites.

A quantitative analysis was carried out for the POM results. The agglomerate area ratio A_R was calculated from the ratio of the area of MWCNT agglomerates A_A to the total investigated area A_0. Furthermore, the dispersion index D was calculated using Eq. (2.1):

$$D = \left(1 - f \frac{A_R}{\varnothing_{\text{vol}}}\right) \times 10 \tag{2.1}$$

Compression Molded Conductive Polymer Composites 21

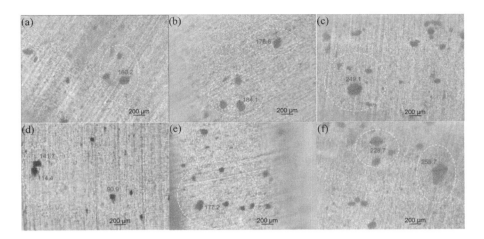

FIGURE 2.2 Optical microscopic images of the extruded (a–c) and compression molded (d–f) samples with 1, 2, and 4 wt% MWCNTs, respectively. MWCNTs, multi-walled carbon nanotubes.

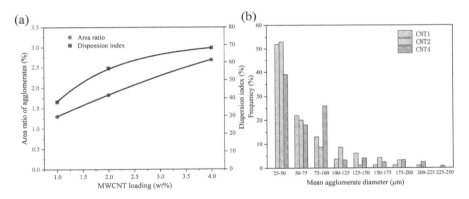

FIGURE 2.3 Area ratio and dispersion index (a) and statistical results on the primary agglomerate size (b) of the compression molded HDPE/MWCNT composites. (In the legend, CNT1 means 1 wt% MWCNTs, and so forth.) HDPE, high-density polyethylene; MWCNTs, multi-walled carbon nanotubes.

where $\varnothing_{vol}$ is the MWCNT volume fraction and f is a factor related to the density of MWCNT agglomerates and was estimated to be 0.25 [2].

The area ratio AR and dispersion index D for the compression molded HDPE/MWCNT nanocomposites are shown in Figure 2.3a. The area ratio clearly represents a linear increase, which means a higher percentage of volume occupied by the nanotube agglomerates in the matrix with increasing MWCNT content [3]. The dispersion index is a more comprehensive method to evaluate the dispersion state of MWCNTs and mixing efficiency because the specific MWCNT loadings are taken into account. In this study, the dispersion index increases when more MWCNTs are added, which can be directly attributed to the decreased ratio of the area ratio AR and the nanotube volume fraction $\varnothing_{vol}$. The increasing dispersion index indicates that further addition of MWCNTs can be dispersed by melt mixing, but the decreased slope with increasing MWCNTs reveals that it is not possible to fully disperse the additional MWCNTs. Consequently, the undispersed MWCNTs mainly exist in the matrix in the form of agglomerates.

A statistical analysis of the size distribution of MWCNT agglomerates was also conducted. The mean diameter distribution of primary agglomerates for the compression molded samples is shown in Figure 2.3b. It can be observed that there are no agglomerates beyond the 200 μm range for the nanocomposites with 1 wt% MWCNTs and very few agglomerates in the range between 150 and 200 μm. For the composites with 4 wt% MWCNTs, a clear increase in the number of agglomerates between 75 and 100 μm in size can be observed, which indicates that many small primary agglomerates grow into bigger agglomerates. Additionally, there are also comparatively more agglomerates between 200 and 225 μm compared to the 2 wt% composites.

2.2.2 Scanning Electron Microscopy

Figure 2.4 shows the morphology of pristine MWCNTs at different magnifications. The individual MWCNTs naturally align themselves into bundles or clusters due to the strong van der Waals force which makes them difficult to disentangle.

Figure 2.5 shows the morphology of extruded and compression molded composite containing 2 wt% MWCNTs. The scanning electron microscope (SEM) image shown in Figure 2.5a is from the extruded sample taken in the extrusion direction. It can be observed in Figure 2.5a that the MWCNTs align along the material flow direction, while the polymer lamellae are perpendicular to the flow direction. A similar alignment of MWCNTs along the flow direction was also reported elsewhere [4,5]. From Figure 2.5b, one can see the individual nanotubes and agglomerates have not formed obvious network structures yet, and the polymer lamellae are randomly aligned after the extruded pellets were compression molded at 150°C for 3 minutes followed by a slow cooling (SC, 20°C/min). A network-like structure nanotube can

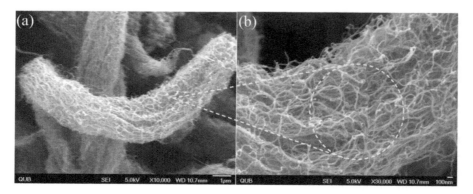

FIGURE 2.4 SEM micrographs of pristine MWCNTs observed at a magnification of (a) ×10,000 and (b) ×30,000, respectively. MWCNTs, multi-walled carbon nanotubes.

Compression Molded Conductive Polymer Composites 23

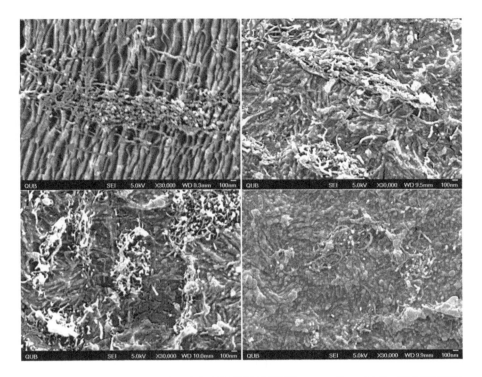

FIGURE 2.5 SEM micrographs of the HDPE/MWCNT composites with 2 wt% MWCNTs: (a) extruded; (b) compression molded at 150°C for 3 minutes followed by SC; (c) compression molded at 200°C for 5 minutes followed by SC; (d) compression molded at 200°C for 5 minutes followed by RC (the agglomerated nanotubes are circled in the micrographs). HDPE, high-density polyethylene; MWCNTs, multi-walled carbon nanotubes.

be observed in Figure 2.5c due to the relaxation of nanotubes after compression molding at 200°C for 5 minutes followed by SC. These individual MWCNTs and secondary agglomerates that can enhance the formation of conductive networks [6] are homogeneously distributed in the HDPE matrix. Figure 2.5d shows the morphology of the composites containing 2 wt% MWCNTs compression molded at 200°C for 5 minutes followed by a rapid cooling (RC, 300°C minutes^{-1}). The polymer lamellae are less clear due to the presence of imperfect crystallites.

2.2.3 Thermal Properties

The thermal properties of the compression molded HDPE and composites with increasing MWCNT loadings were investigated using differential scanning calorimetry (DSC). The crystallinity (X_c^{2nd}) and melting temperature (T_m^{2nd}) in the second heating stage and crystallization temperature (T_c) are shown in Table 2.1 [7].

It can be seen in Table 2.1 that the crystallization temperature increases by about 2°C with the addition of MWCNTs, indicating that MWCNTs are acting as nucleation sites [8]. However, the addition of MWCNTs does not show an effect on the melting temperature of the material; thus, the temperature difference between melting

TABLE 2.1
Crystallization and Melting Data of HDPE/MWCNT Composites Obtained by DSC

Sample	CNT Loading (wt%)	X_c^{2nd} (%)	T_m^{2nd} (°C)	T_c (°C)
HDPE	0	70.4 ± 1.3	134.1 ± 0.1	114.0 ± 0.2
CNT1	1	70.4 ± 2.7	134.5 ± 0.1	115.7 ± 0.1
CNT2	2	69.2 ± 2.2	134.6 ± 0.4	115.7 ± 0.5
CNT4	4	70.7 ± 2.4	134.3 ± 0.2	116.0 ± 0.2
CNT6	6	68.8 ± 2.0	134.0 ± 0.2	116.2 ± 0.1
CNT8	8	72.6 ± 1.9	133.9 ± 0.4	116.2 ± 0.4
CNT10	10	67.1 ± 5.7	133.6 ± 0.2	116.1 ± 0.1

HDPE, high-density polyethylene; MWCNT, multi-walled carbon nanotube.

temperature and crystallization temperature is reduced. This is significant for the adjustment of relevant processing windows in industrial production.

CNTs may have a complex effect on the crystallinity of the polymer matrix. On the one hand, CNTs may function as heterogeneous nucleating agents for polymer crystallization, while on the other hand, they may hinder molecular mobility during crystallization and hence reduce crystallinity [9]. Therefore, different results for the crystallinity of polymer nanocomposites have been reported in the previous literature. A trend toward a slight decrease in the crystallinity of PE/CNT composites was observed as the content of nanotubes increases in [10,11]. However, in other studies [3,8,12], there was no significant difference in the crystallinity of the PE/CNT composites. In this study, there appears to be no significant effect of MWCNTs on the crystallinity in the 2nd heating stage as measured by DSC (Table 2.1).

In order to investigate the effect of cooling rate on the thermal properties of unfilled HDPE and HDPE/MWCNT composites, the crystallinity and melting temperature in the first heating stage of compression molded samples with different cooling rates were investigated by DSC, as shown in Figure 2.6. The results show that the rapidly cooled samples exhibit a lower melting temperature, which can be attributed to the smaller crystallites formed. The crystallinity in the first heating stage decreases after RC due to the suppression effect of RC on polymer mobility [13].

2.2.4 Tensile Properties

The compression molded samples were tensile tested to investigate the effect of the addition of MWCNTs on the mechanical properties of the polymer matrix, as shown in Table 2.2. The Young's modulus (E) is significantly improved with the presence of MWCNTs. However, the increase in modulus is not a linear relationship with MWCNT loading. There rate of modulus increase reduces at higher MWCNT loadings (>6 wt%), indicating that the nanotubes are less effective in reinforcing at higher loading levels. Due to the presence of more particle aggregates (lower effective particle aspect ratio), the stress transfer efficiency between polymer matrix and MWCNT is low under higher

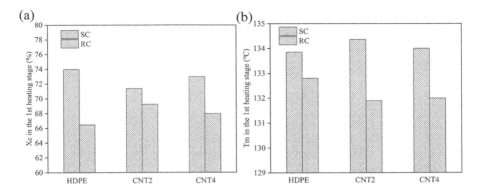

FIGURE 2.6 Crystallinity (a) and melting temperature (b) in the first heating stage of compression molded samples with different cooling rates.

TABLE 2.2
Effect of MWCNTs on the Tensile Properties of HDPE/MWCNT Composites Compression Molded at 200°C for 5 minutes Followed by SC

Sample	CNT Loading (wt%)	E (MPa)	ΔE (%)	σ_b (MPa)	$\Delta\sigma_b$ (%)	ε_b (%)	$\Delta\varepsilon_b$ (%)
HDPE	0	900.6 ± 61.1	-	27.9 ± 3.0	-	907.0 ± 58.0	-
CNT1	1	1124.5 ± 24.0	+24.9	18.5 ± 0.4	−33.6	111.7 ± 39.8	−87.7
CNT2	2	1278.9 ± 27.4	+42.0	17.1 ± 0.7	−38.5	70.0 ± 16.0	−92.3
CNT4	4	1905.8 ± 149.3	+111.6	16.6 ± 0.3	−40.5	46.9 ± 8.8	−94.8
CNT6	6	2364.7 ± 227.2	+162.6	14.7 ± 0.9	−47.3	29.5 ± 7.6	−96.7
CNT8	8	2372.6 ± 243.5	+163.4	18.2 ± 1.5	−34.6	20.4 ± 2.2	−97.8
CNT10	10	2444.0 ± 108.6	+171.4	19.9 ± 1.5	−28.6	23.5 ± 4.2	−97.4

HDPE, high-density polyethylene; MWCNT, multi-walled carbon nanotube; SC, slow cooling.

MWCNT load. The improvement in E is about 110%–160% with the addition of 4–6 wt% MWCNTs. This improvement is much higher than the previously reported improvement (40%–70%) at 3–6 wt% MWCNTs in studies on HDPE/MWCNT composites [4,5,14]. The greatest improvement of 171.4% in E is obtained at a MWCNT content of 10 wt%.

MWCNT inclusion has a negative effect on the stress at break (σ_b) and strain at break (ε_b) of all the composites. The σ_b and ε_b of the composites containing 1~10 wt% MWCNTs decreases by 29%–47% and 88%–97%, respectively, in this case (Table 2.2). This deterioration can be attributed to the presence of MWCNT agglomerates in the nanocomposites. Since failure initiates at the weakest point in a material, CNT agglomerates may work as initiation sites for the propagation and extension of cracks and accelerate the final breakage [5]. Chrissafis et al. [15] also found a significant decrease in elongation (decreased by 93%) and tensile strength (decreased by 55%) with the addition of 2.5 wt% of MWCNTs to HDPE.

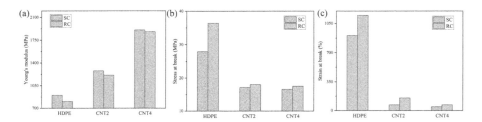

FIGURE 2.7 Effect of cooling rate on the modulus (a), stress at break (b), and strain at break (c) of HDPE/MWCNT composites. HDPE, high-density polyethylene; MWCNT, multi-walled carbon nanotube.

The tensile properties of samples compression molded at 200°C for 5 minutes followed by RC or SC were compared, as shown in Figure 2.7. RC led to a slight decrease of 11.6% in modulus for the unfilled HDPE samples due to a lower crystallinity level and imperfect crystallites, but it improved the stress at break and strain at break by 30.4% and 26.6%, indicating that the toughness of rapidly cooled HDPE samples is enhanced. However, the effect of cooling rate on the tensile properties of HDPE composites containing 2 and 4 wt% MWCNTs is not significant considering the experimental errors.

2.2.5 ELECTRICAL PROPERTIES

The electrical properties of HDPE/MWCNT nanocomposites were characterized by measuring the volume resistivity of extruded and compression molded samples, as shown in Figure 2.8. The nanocomposites were compression molded with different heating temperatures (150°C or 200°C) and holding times (3 or 5 minutes) followed by SC. According to the traditional percolation theory of highly dispersed conductive additives in an isolating polymer matrix, conductive fillers can form conductive networks in the bulk of the matrix, thus causing a decrease in the resistivity when the content of the conductive additive reaches the percolation threshold.

The resistivity of extruded pellets with 2 wt% MWCNTs is still beyond the full scale of the multimeter showing a high resistivity (>10^8 Ω), but it decreases markedly when the MWCNT content reaches 4 wt%. Interestingly, the resistivity of all the compression molded composites is lower than that of the extruded pellets, even though the resistivity measurement is in the longitudinal direction of the extruded pellet and electron transport is facilitated in the direction parallel to nanotube alignment [16,17]. The resistivity of compression molded composites decreases with increasing heating temperature and holding time. Zhang [18] reported similar behavior for hot-pressed and extruded TPU/CNT composites, in which an extremely low percolation threshold of 0.13 wt% was achieved in hot-pressed composite film samples, whereas a much higher CNT concentration (3–4 wt%) was needed to form a conductive network in the extruded composite strands. The difference in conductivity between the extruded and compression molded samples may be attributed to the difference in alignment of the CNTs in the samples with alignment being greater in the extruded pellets and thus having a lower potential to form a conductive network [19].

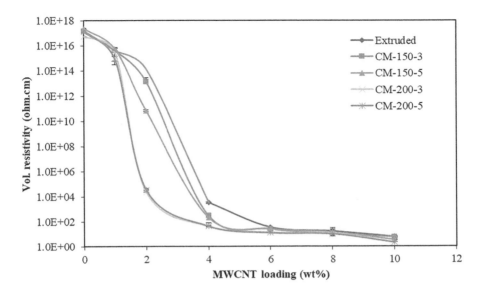

FIGURE 2.8 Volume resistivity of extruded HDPE/MWCNT nanocomposites and compression molded HDPE/MWCNT nanocomposites with different heating temperatures and holding time followed by SC as a function of MWCNT loading. HDPE, high-density polyethylene; MWCNT, multi-walled carbon nanotube.

The conductive network in the HDPE/MWCNT composites was further analyzed according to the scaling law (Eq. 2.2) [4] of classical percolation theory.

$$\rho \propto (\varnothing - \varnothing_t)^{-t} \qquad (2.2)$$

where ρ is the volume resistivity of the composite, $\varnothing$ is the filler weight fraction, $\varnothing_t$ is the critical concentration, and t is the critical exponent, which reflects the system dimensionality of the composite. It follows a power-law dependence of approximately 1.6–2 in a three-dimensional system and 1–1.3 in a two-dimensional system [20]. The critical concentration $\varnothing_t$ and the critical exponent t of the composites were listed in Table 2.3.

The results reveal a critical concentration of 3.9 wt% and a critical exponent of 1.3 for the extruded composites. This low critical exponent indicates that the extruded composites generally follow a two-dimensional model due to the restriction for electron hopping between the MWCNTs in the transverse direction. The critical concentration decreases and the critical exponent increases with increasing heating temperature and holding time. The composite compression molded at a heating temperature of 200°C and a holding time of 5 minutes followed by SC shows a critical concentration of 1.9 wt% and a critical exponent of 1.9. It is evident that the decrease in critical concentration means an improvement in conductivity during the compression molding process, and the increase in critical exponent reveals the steady transformation from a two-dimensional system to a three-dimensional system after compression molding.

TABLE 2.3
Critical Concentration $\varnothing_t$ and Critical Exponent t for the HDPE/MWCNT Composites with Different Heating Temperatures and Holding Time Followed by SC

Parameters	CM-150-3	CM-150-5	CM-200-3	CM-200-5	Extruded
$\varnothing_t$ (wt%)	3.5	3.3	2.0	1.9	3.9
t	1.5	1.5	1.7	1.9	1.3

HDPE, high-density polyethylene; MWCNT, multi-walled carbon nanotube; SC, slow cooling.

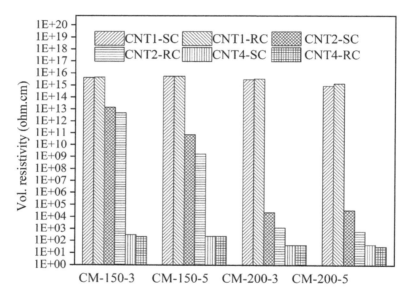

FIGURE 2.9 Resistivity variations of the HDPE/MWCNT composites with MWCNT loadings in the region of the electrical percolation threshold under different compression molding conditions. HDPE, high-density polyethylene; MWCNT, multi-walled carbon nanotube

The compression molded composites containing only 2 wt% MWCNTs exhibit a lower resistivity (~3×10^4 Ω cm) in this study, comparing to the resistivity (~1×10^8 Ω cm) of compression molded composites with 3 wt% MWCNTs in Verge's [21] investigation, in which the same HDPE matrix (HTA-108) and MWCNTs (NC7000) were produced via melt mixing. The significant improvement in the electrical conductivity of the composites produced in this study is likely to be due to more uniformly dispersed individual nanotubes and secondary nanotube agglomerates as a result of more effective melt mixing.

It can be seen in Figure 2.8 that the electrical properties of composites with a MWCNT loading close to the percolation threshold are significantly influenced by compression molding conditions. Details of the resistivity of the nanocomposites containing 1, 2, and 4 wt% MWCNTs are shown in Figure 2.9. Also, the volume electrical resistivity of the composites produced under SC and RC is compared. It can be seen clearly in Figure 2.9 that the resistivity of the composite containing 2 wt% MWCNTs is significantly influenced by the heating temperature and cooling rate. The resistivity of the composite containing 2 wt% MWCNTs decreased by six to nine orders of magnitude when the heating temperature increased from 150°C to 200°C. Interestingly, the electrical resistivity of the rapidly cooled composite with 2 wt% MWCNTs is one to two orders lower than that of the slowly cooled composites with the same MWCNT loading. This may be due to the lower crystallinity and smaller crystallites facilitating the formation of conductive pathways [22,23]. The effect of cooling rate on resistivity is much less significant for the composite with 4 wt% MWCNTs as a result of the abundant conductive networks formed in the HDPE matrix. This result indicates that cooling rate can be a significant parameter in influencing electrical conductivity when operating in the region of the percolation threshold.

2.3 UNARY CARBON NANOFILLER-REINFORCED COMPOSITES

In this section, a comparative study of unary carbon nanofiller-filled composites containing 4 wt% nanofillers was conducted to assess the performance and reinforcement effect of carbon nanofillers with different dimensions. All the samples investigated in this comparative study were compression molded under the same molding conditions (heating temperature: 200°C, holding time: 5 minutes, SC).

2.3.1 SCANNING ELECTRON MICROSCOPY

The morphology of pristine GNPs and CBs was observed by SEM. It can be seen in Figure 2.10 that the GNPs and CBs present typical two-dimensional and zero-dimensional shape characteristics, respectively. Similar to CNTs, these GNPs and CBs also tend to aggregate due to a strong van der Waals force.

Figure 2.11 shows the morphology of compression molded unary carbon nanofiller composites with 4 wt% MWCNTs, GNPs, or CB respectively. It can be observed in Figure 2.11 that these individual and agglomerated nanofillers are distributed uniformly in the polymer matrix. The stacked layers of the GNPs are not broken into single layers by the melt mixing due to the strong van der Waals force and the additional interplanar π–π interactions between the individual graphene sheets [24,25] as shown in Figure 2.11c. The net-like CB clusters in Figure 2.11d may facilitate the formation of conductive pathways [26,27].

2.3.2 THERMAL PROPERTIES

The crystallization and melting behavior of the unary carbon nanofiller-reinforced composites with 4 wt% nanofillers were investigated using DSC. The crystallinity (X_c^{2nd}) and melting temperature (T_m^{2nd}) in the second heating stage and crystallization temperature (T_c) are shown in Table 2.4. It can be seen in Table 2.4 that the crystallinity and melting temperature are barely changed, regardless of the type of carbon nanofillers. The crystallization temperature increased by 1.5°C–2°C due to the nucleation effect of carbon nanofillers [8,28]. A higher T_c for the HDPE/MWCNT composite indicates that the one-dimensional nanotubes may more effectively facilitate the nucleation of HDPE, while the CBs show the weakest nucleation effect.

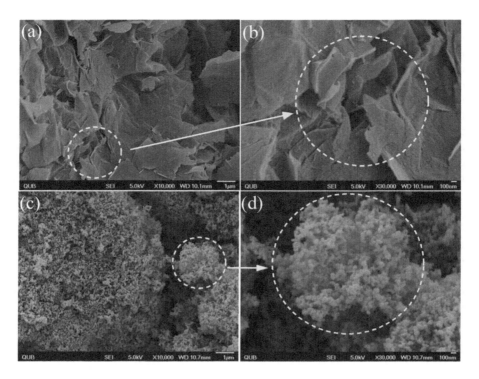

FIGURE 2.10 SEM micrographs of pristine GNPs and CBs: (a) GNPs, ×10,000; (b) GNPs, ×30,000; (c) CBs, ×10,000; and (d) CBs, ×30,000.

2.3.3 Tensile Properties

The tensile properties of the compression molded unary carbon nanofiller composites with 4 wt% nanofillers are compared in Table 2.5. The nanocomposites filled with GNPs, which have the highest aspect ratio, exhibit the highest reinforcement (125.5%) in modulus as a result of effective stress transfer between the GNPs and HDPE matrix. The modulus only increased by 29.7% for the HDPE/CB composite due to the low aspect ratio of CB.

The experimental values of modulus were compared with predicted values using the randomly aligned Halpin–Tsai model $\left(\frac{E_r}{E_m} = \frac{3}{8}\left[\frac{1+\xi\eta_L\varphi_f}{1-\eta_L\varphi_f} \right] + \frac{5}{8}\left[\frac{1+2\eta_T\varphi_f}{1-\eta_T\varphi_f} \right] \right)$ described, as shown in Figure 2.12. The experimental values agree very well with the predicted values for all the unary carbon nanofiller-reinforced composites. This agreement indicates a uniform dispersion of particles in the composite as a result of melt mixing. However, it can be observed in Figure 2.12 that the experimental modulus is 19% higher than the predicted value for the HDPE/CB composite. This may be attributed to the fact that the high structure CB used with a high surface roughness (high surface area 1,400 m^2/g) is not of perfect spherical structure; thus, their practical aspect ratio is higher than the aspect ratio of 1, which was used in the prediction model. Also, a high surface area of dispersed phase may be presented to the continuous phase with consequences for surface adsorption and immobilization [29].

Compression Molded Conductive Polymer Composites

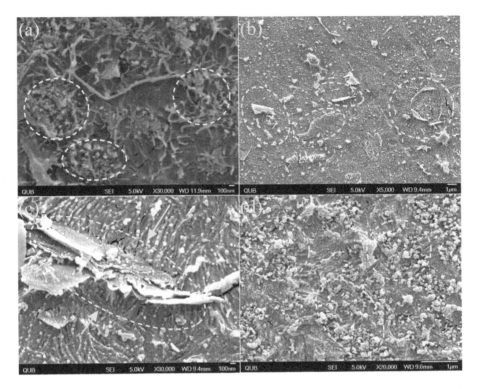

FIGURE 2.11 SEM micrographs of unary carbon nanofiller-reinforced composites with 4 wt% nanofillers: (a) HDPE/MWCNT, ×30,000; (b) HDPE/GNP, ×5,000; (c) HDPE/GNP, ×30,000; and (d) HDPE/CB, ×20,000 (the agglomerated nanofillers are circled, and the individual nanofillers are indicated by arrows in the micrographs). HDPE, high-density polyethylene; MWCNT, multi-walled carbon nanotube.

TABLE 2.4
Crystallization and Melting Data of Unary Carbon Nanofiller-Reinforced Composites Obtained by DSC

Sample	Nanofiller Loading	X_c^{2nd} (%)	T_m^{2nd} (°C)	T_c (°C)
HDPE	0 wt%	70.4 ± 1.3	134.1 ± 0.1	114.0 ± 0.2
CNT4	4 wt% CNTs	70.7 ± 2.4	134.3 ± 0.2	116.0 ± 0.2
GNP4	4 wt% GNPs	70.3 ± 2.7	134.5 ± 0.2	115.6 ± 0.2
CB4	4 wt% CB	68.6 ± 0.8	134.6 ± 0.4	114.5 ± 0.3

DSC, differential scanning calorimetry; HDPE, high-density polyethylene; CNT, carbon nanotube; GNP, graphene nanoplatelet; CB, carbon black.

TABLE 2.5
Tensile Properties of the Unary Carbon Nanofiller-Reinforced Composites with 4 wt% Nanofillers

Sample	Nanofiller Loading	E (MPa)	ΔE (%)	σ_b (MPa)	$\Delta \sigma_b$ (%)	ε_b (%)	$\Delta \varepsilon_b$ (%)
HDPE	0 wt%	900.6 ± 61.1	-	27.9 ± 3.0	-	907.0 ± 58.0	-
CNT4	4 wt% CNTs	1905.8 ± 149.3	+111.6	16.6 ± 0.3	−40.5	46.9 ± 8.8	−94.8
GNP4	4 wt% GNPs	2031.2 ± 96.3	+125.5	23.6 ± 1.2	−15.4	12.5 ± 2.1	−98.6
CB4	4 wt% CB	1167.8 ± 99.5	+29.7	16.4 ± 0.8	−40.1	27.2 ± 10.4	−97.0

DSC, differential scanning calorimetry; HDPE, high-density polyethylene; CNT, carbon nanotube; GNP, graphene nanoplatelet; CB, carbon black

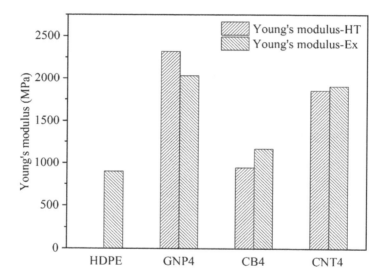

FIGURE 2.12 Experimental and predicted modulus of the unary carbon nanofiller-reinforced composites with 4 wt% nanofillers.

It can be seen in Table 2.5 that the addition of carbon nanofillers has a negative effect on the stress at break and strain at break due to the presence of agglomerates, regardless of the dimensionality of the nanofiller. The stress at break of CB and MWCNTs filled HDPE composites decreases by about 40%. The stress at break of HDPE/GNP composites decreases by only 15.4%, while its elongation decreases significantly by 98.6%.

2.3.4 ELECTRICAL PROPERTIES

The volume resistivity of unary carbon nanofiller-reinforced composites is shown in Figure 2.13. The HDPE/GNP composite exhibits a high resistivity, indicating insufficient conductive pathways in the insulating HDPE matrix. This poor enhancement

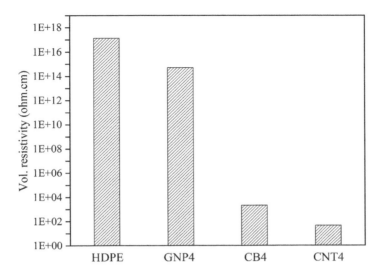

FIGURE 2.13 Volume resistivity of the unary carbon nanofiller-reinforced composites with 4 wt% nanofillers.

with the highest aspect ratio particles may be due to the difficulty in achieving interlacing of the particles to produce a conductive network [20]. In addition, the commercial GNPs used in this study may contain some unconductive graphite oxide [24]. The resistivity of HDPE decreases significantly by 14 and 16 orders of magnitude with the addition of 4 wt% CB or MWCNTs, respectivly. The one-dimensional MWCNTs are the most efficient in terms of generating conductive networks, which are likely to be due to the ease with which one-dimensional particles can entangle and form networks. Altough the CB has a low aspect ratio, high structure CB tends to agglomerate in branched clusters that are beneficaial in the formation of conductive pathways [26,27]. Therefore, it can be seen in this study that the relative effectiveness of enhancement in electrical property is as follows: GNPs < CB < MWCNTs, at the same loading.

2.4 BINARY CARBON NANOFILLER-REINFORCED COMPOSITES

In this section, the properties of compression molded binary carbon nanofiller-filled composites with 4 wt% nanofillers are investigated. There are three ratios of weight fraction between the two types of carbon nanofillers: 3:1, 2:2, and 1:3. Here all the samples were compression molded at 200°C for 5 minutes followed by SC.

2.4.1 Scanning Electron Microscopy

Figure 2.14 shows the morphology of compression molded binary carbon nanofiller-reinforced composites with 4 wt% nanofillers in total (the ratio between the two types of nanofillers is 2:2). It can be observed in Figure 2.14 that these individual and

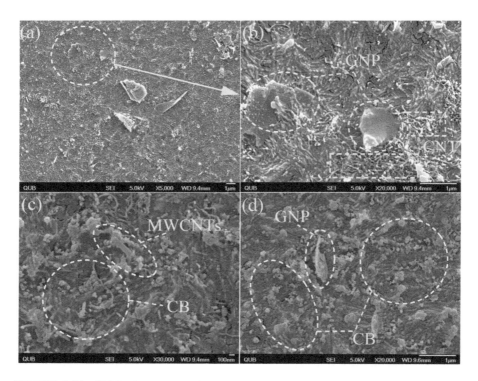

FIGURE 2.14 SEM micrographs of binary carbon nanofiller-reinforced composites with 4 wt% nanofillers: (a and b) HDPE/GNP/MWCNT composite with 2 wt% GNPs and 2 wt% MWCNTs, ×5,000 and ×20,000, respectively; (c) HDPE/CB/MWCNT composite with 2 wt% CB and 2 wt% MWCNTs, ×30,000; and (d) HDPE/GNP/CB composite with 2 wt% GNPs and 2 wt% CB, ×20,000. HDPE, high-density polyethylene; MWCNT, multi-walled carbon nanotube.

agglomerated nanofillers are distributed uniformly in the HDPE matrix. The GNPs and CB are bridged by the MWCNTs in Figure 2.14b and c respectively, and the GNPs are surrounded by the CB in Figure 2.14d [30,31].

2.4.2 Thermal Properties

The thermal properties of the compression molded HDPE/GNP/MWCNT, HDPE/CB/MWCNT, and HDPE/GNP/CB binary carbon nanofiller-reinforced composites with 4 wt% nanofillers were investigated using DSC. The crystallinity (X_c^{2nd}) and melting temperature (T_m^{2nd}) in the second heating stage and crystallization temperature (T_c) are shown in Table 2.6. As expected, it can be seen in Table 2.6 that the crystallinity and melting temperature are not affected by the addition of a combination of two types of carbon nanofillers. The crystallization temperature of the binary carbon nanofiller-reinforced composites increases by about 1.5°C–2°C due to the nucleation effect of carbon nanofillers [8,28]. The unary and binary carbon nanofillers exhibit very similar effects on the thermal properties of composites.

TABLE 2.6
Crystallization and Melting Data of Binary Carbon Nanofiller-Reinforced Composites Obtained by DSC

Sample	Nanofiller Loading	X_c^{2nd} (%)	T_m^{2nd} (°C)	T_c (°C)
HDPE	0 wt%	70.4 ± 1.3	134.1 ± 0.1	114.0 ± 0.2
GNP3CNT1	3 wt% GNPs + 1 wt% CNTs	70.7 ± 1.4	134.5 ± 0.1	116.1 ± 0.2
GNP2CNT2	2 wt% GNPs + 2 wt% CNTs	72.9 ± 2.8	134.0 ± 0.5	116.2 ± 0.2
GNP1CNT3	1 wt% GNPs + 3 wt% CNTs	69.7 ± 2.1	134.2 ± 0.3	116.3 ± 0.3
CB3CNT1	3 wt% CB + 1 wt% CNTs	70.9 ± 2.5	134.4 ± 0.1	115.5 ± 0.1
CB2CNT2	2 wt% CB + 2 wt% CNTs	70.6 ± 3.0	134.5 ± 0.2	115.7 ± 0.2
CB1CNT3	1 wt% CB + 3 wt% CNTs	69.6 ± 1.8	134.5 ± 0.2	115.8 ± 0.2
GNP3CB1	3 wt% GNPs + 1 wt% CB	70.8 ± 3.4	134.6 ± 0.2	115.6 ± 0.3
GNP2CB2	2 wt% GNPs + 2 wt% CB	70.5 ± 3.0	134.5 ± 0.2	115.6 ± 0.4
GNP1CB3	1 wt% GNPs + 3 wt% CB	70.0 ± 1.0	134.4 ± 0.1	115.4 ± 0.2

DSC, differential scanning calorimetry; HDPE, high-density polyethylene; CNT, carbon nanotube; GNP, graphene nanoplatelet; CB, carbon black

2.4.3 TENSILE PROPERTIES

The tensile properties of compression molded binary carbon nanofillers-reinforced composites with 4 wt% nanofillers were investigated, as shown in Table 2.7. It can be observed in Table 2.7 that the elastic modulus of the binary carbon nanofillers-reinforced composites is improved significantly with the addition of carbon nanofillers. For the HDPE/CB/MWCNT composites, the modulus increases with increasing MWCNT loadings. Similarly, the modulus of the HDPE/GNP/CB composites increases with increasing loading of GNPs. The increase in modulus (over 114%) for all the HDPE/GNP/MWCNT composites is higher than that for the HDPE/CB/MWCNT and HDPE/GNP/CB composites due to the high aspect ratios of both GNPs and MWCNTs.

The experimental values of modulus of the binary carbon nanofillers-reinforced composites were compared with predicted values using the Halpin–Tsai method described, as shown in Figure 2.15. Interestingly, the experimental modulus of the HDPE/GNP/MWCNT composite containing 2 wt% GNPs and 2 wt% MWCNTs is higher than the predicted modulus and the modulus of the composite containing 4 wt% GNPs (Table 2.5), which may be attributed to a synergistic effect arising from nanofiller–nanofiller network structures [32–34], formed in the system enhancing the stress transfer between the polymer and nanofillers. The network structure is generated due to numerous entanglements between the polymer chains and nanofillers, and the polymer chains act as a "bridge" via polymer wrapping of carbon nanofillers. A relatively weaker synergistic effect can be observed for the HDPE/GNP/MWCNT composite with 1 wt% GNPs and 3 wt% MWCNTs, which may be attributed to the decreased GNP content (GNPs exhibit the most efficient reinforcement in modulus). A similar synergistic effect is also observed in the HDPE/CB/MWCNT composites. All the experimental values of modulus of the HDPE/CB/MWCNT composites are

TABLE 2.7
Tensile Properties of the Binary Carbon Nanofiller-Reinforced Composites with 4 wt% Nanofillers

Sample	Nanofiller Loading	E (MPa)	ΔE (%)	σ_b (MPa)	$\Delta\sigma_b$ (%)	ε_b (%)	$\Delta\varepsilon_b$ (%)
HDPE	0 wt%	900.6 ± 61.1	-	27.9 ± 3.0	-	907.0 ± 58.0	-
GNP3CNT1	3 wt% GNPs + 1 wt% CNTs	1930.6 ± 118.8	+114.4	21.9 ± 0.9	−21.5	17.1 ± 3.0	−98.1
GNP2CNT2	2 wt% GNPs + 2 wt% CNTs	2403.3 ± 100.3	+166.9	25.2 ± 1.9	−9.7	11.8 ± 2.8	−98.7
GNP1CNT3	1 wt% GNPs + 3 wt% CNTs	2044.6 ± 43.6	+127.0	20.4 ± 0.9	−26.9	43.6 ± 8.7	−95.2
CB3CNT1	3 wt% CB + 1 wt% CNTs	1706.6 ± 48.8	+89.5	17.0 ± 0.6	−39.1	43.1 ± 11.47	−95.2
CB2CNT2	2 wt% CB + 2 wt% CNTs	1787.8 ± 71.9	+98.5	16.8 ± 0.6	−39.8	39.5 ± 11.6	−95.6
CB1CNT3	1 wt% CB + 3 wt% CNTs	1791.3 ± 57.1	+98.9	16.6 ± 0.2	−40.5	42.6 ± 16.1	−95.3
GNP3CB1	3 wt% GNPs + 1 wt% CB	1901.9 ± 116.6	+111.2	22.4 ± 1.1	−19.7	13.1 ± 4.0	−98.6
GNP2CB2	2 wt% GNPs + 2 wt% CB	1639.4 ± 51.2	+82.0	21.6 ± 1.5	−22.5	17.4 ± 5.2	−98.1
GNP1CB3	1 wt% GNPs + 3 wt% CB	1381.2 ± 128.4	+53.4	17.3 ± 0.6	−38.0	31.1 ± 13.2	−96.6

DSC, differential scanning calorimetry; HDPE, high-density polyethylene; CNT, carbon nanotube; GNP, graphene nanoplatelet; CB, carbon black.

higher than the predicted values. The modulus of the composite containing 3 wt% CB and 1 wt% MWCNTs significantly increased by 46% compared to the composite containing 4 wt% CB. However, it is not evident for the HDPE/GNP/CB composites due to the lack of nanofiller–nanofiller network structures. A schematic diagram of nanofiller–nanofiller network structure is shown in Figure 2.16.

It can be seen in Table 2.7 that the addition of binary carbon nanofillers still exhibits a negative effect on the stress at break and strain at break. The stress at break of the HDPE/GNP/MWCNT composite containing 2 wt% GNPs and 2 wt% MWCNTs decreased by only 9.7% compared to the unfilled HDPE, but its strain at break decreased significantly by 98.7%.

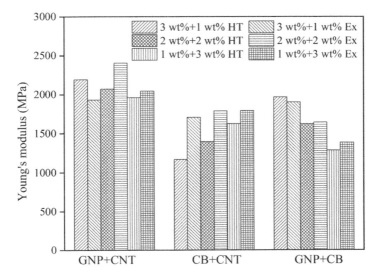

FIGURE 2.15 Experimental and predicted Young's modulus of the binary carbon nanofiller-reinforced composites.

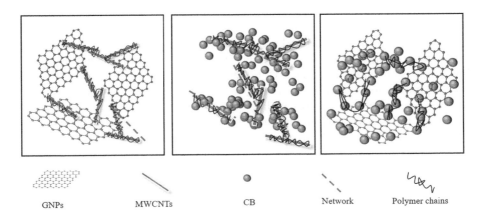

FIGURE 2.16 Schematic diagram of the nanofiller–nanofiller network structure formed in HDPE/GNP/MWCNT and HDPE/CB/MWCNT composites. HDPE, high-density polyethylene; MWCNT, multi-walled carbon nanotube.

2.4.4 Electrical Properties

The volume resistivity of binary carbon nanofiller-reinforced composites is shown in Figure 2.17. It can be observed in Figure 2.17 that the resistivity of all the HDPE/CB/MWCNT composites with 4 wt% nanofillers significantly decreased by 15 orders of magnitude compared to that of unfilled HDPE (1.3×10^{17} Ω cm), indicating abundant conductive pathways formed with the addition of MWCNTs and high structure CB. For the HDPE/GNP/MWCNT and HDPE/GNP/CB composites, the resistivity

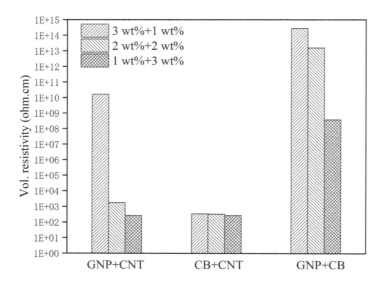

FIGURE 2.17 Volume resistivity of the binary carbon nanofiller-reinforced composites.

decreases with increasing loadings of MWCNTs or CB respectively, while it is clear that the 1D MWCNTs with higher aspect ratio are more efficient than the spherical CB in terms of generating conductive networks in the binary carbon nanofiller-reinforced systems. By way of example, compared to the composite containing 4 wt% GNPs, the resistivity decreased by 11 orders of magnitude for the HDPE/GNP/MWCNT composite with 2 wt% GNPs and 2 wt% MWCNTs, while it only decreased by 1 order of magnitude for the HDPE/GNP/CB composite with 2 wt% GNPs and 2 wt% CB. In this study, the resistivity of binary carbon nanofiller-reinforced composites is between that of corresponding unary carbon nanofiller composites shown in Figure 2.13 in Section 2.3.4, thus there is no evident synergistic effect in the resistivity of binary carbon nanofillers-reinforced composites.

2.5 CONCLUSIONS

First, the molded HDPE/MWCNT composites containing 1–10 wt% MWCNTs prepared under different molding conditions were systematically studied. The results show that the addition of MWCNTs has no significant effect on the crystallinity, average grain size, and melting temperature, but RC during the molding process will slightly reduce the crystallinity, average grain size, and melting temperature. After adding 4–6 wt% MWCNT, the Young's modulus of the molded composites was significantly increased by 110–160%. When the MWCNT load is <6 wt%, the predicted modulus is in good agreement with the experimental results. However, due to the existence of MWCNT aggregates, the fracture stress and fracture strain of composites containing 1–10 wt% MWCNTs are reduced by 29%–47% and 88%–97%, respectively. The volume resistivity test shows that the resistivity of all molded HDPE/MWCNT composites is lower than that of extruded particles. According to the scaling law of classical seepage theory, in the process of compression molding, with the increase of heating temperature and holding time, the critical concentration decreases and the critical index increases. This is attributed to the enhanced relaxation of MWCNTs. Under the conditions of heating temperature of 200°C, holding time of 5 minutes and SC, the critical concentration of the molded composite is 1.9 wt% and the critical index is 1.9. Interestingly, the resistivity of fast cooling composites containing 2 wt% MWCNT is one to two orders of magnitude lower than that of SC composites with the same MWCNT load. This may be because lower crystallinity and smaller crystals promote the formation of conductive pathways.

Subsequently, a comparative study of compression molded unary carbon nanofiller composites with 4 wt% nanofillers was carried out in order to assess the reinforcement effects of carbon nanofillers with different dimensions. The crystallinity and average crystallite sizes of the unary carbon nanofiller composites are not influenced significantly by nanocarbon content regardless of the dimensionality of the carbon nanofiller. The crystallization temperature increased by 1.5°C–2°C due to the nucleation effect of carbon nanofillers. Resistivity tests indicate that the relative effectiveness of conductive networks is as follows: GNPs < CB < MWCNTs, at the same loading. Nanocomposites filled with GNPs, which have the highest aspect ratio, exhibit the highest reinforcement (125.5%) in modulus. However, the modulus only increased by 29.7% for the HDPE/CB composite due to the low aspect ratio of CB. The experimental modulus agrees well with the predicted values for all the unary carbon nanofiller-reinforced composites. The addition of carbon nanofillers has a negative effect on the stress at break and strain at break, regardless of the dimensionality of the nanofiller.

Third, the properties of compression molded binary carbon nanofiller composites were investigated to determine any possible synergistic effects arising from a combination of two different types of carbon nanofiller. It was found that conductive networks were formed in all the HDPE/CB/MWCNT composites. For the HDPE/GNP/MWCNT and HDPE/GNP/CB composites, the conductive networks were enhanced with increasing MWCNT or CB loadings, respectively, while increasing MWCNT content in the HDPE/GNP/MWCNT composites more facilitates the formation of conductive networks compared to increasing CB content in the HDPE/GNP/CB composites. The experimental modulus of the HDPE/GNP/MWCNT composite containing 2 wt% GNPs and 2 wt% MWCNTs is higher than the predicted modulus, which may be attributed to nanofiller–nanofiller network structures formed in the system enhancing the stress transfer between the polymer and nanofillers. A similar synergistic effect is also observed in the HDPE/CB/MWCNT composites, but it is not evident for the HDPE/GNP/CB composites due to the lack of nanofiller–nanofiller network structures. The addition of binary carbon nanofillers exhibits a clear negative effect on the stress at break and strain at break similar to what was observed with the unary carbon nanofiller-reinforced system.

REFERENCES

1. Xiang, D, Harkin-Jones E, Linton D, Martin P. Structure, mechanical, and electrical properties of high-density polyethylene/multi-walled carbon nanotube composites processed by compression molding and blown film extrusion. *J Appl Polym Sci.* 2015;132(42):42665.

2. Villmow T, Pötschke P, Pegel S, Häussler L, Kretzschmar B. Influence of twin-screw extrusion conditions on the dispersion of multi-walled carbon nanotubes in a poly(lactic acid) matrix. *Polymer.* 2008;49(16):3500–9.
3. Chakraborty S, Pionteck J, Krause B, Banerjee S, Voit B. Influence of different carbon nanotubes on the electrical and mechanical properties of melt mixed poly(ether sulfone)-multi walled carbon nanotube composites. *Compos Sci Technol.* 2012;72(15):1933–40.
4. Nanni F, Mayoral BL, Madau F, Montesperelli G, McNally T. Effect of MWCNT alignment on mechanical and self-monitoring properties of extruded PET-MWCNT nanocomposites. *Compos Sci Technol.* 2012;72(10):1140–6.
5. Pollanen M, Pirinen S, Suvanto M, Pakkanen TT. Influence of carbon nanotube-polymeric compatibilizer masterbatches on morphological, thermal, mechanical, and tribological properties of polyethylene. *Compos Sci Technol.* 2011;71(10):1353–60.
6. Socher R, Krause B, Muller MT, Boldt R, Potschke P. The influence of matrix viscosity on MWCNT dispersion and electrical properties in different thermoplastic nanocomposites. *Polymer.* 2012;53(2):495–504.
7. Xiang D, Wang L, Tang Y, Hill CJ, Chen B, Harkin-Jones E. Reinforcement effect and synergy of carbon nanofillers with different dimensions in high density polyethylene based nanocomposites. *Int J Mater Res.* 2017;108(4):322–34.
8. Haggenmueller R, Fischer J, Winey K. Single wall carbon nanotube/polyethylene nanocomposites: Nucleating and templating polyethylene crystallites. *Macromolecules.* 2006;39:2964–71.
9. Xu DH, Wang ZG. Role of multi-wall carbon nanotube network in composites to crystallization of isotactic polypropylene matrix. *Polymer.* 2008;49(1):330–8.
10. Linares A, Canalda J, Cagiao ME, García-Gutiérrez MC, Nogales A, Martin-Gullon I, et al. Broad-band electrical conductivity of high density polyethylene nanocomposites with carbon nanoadditives: Multiwall carbon nanotubes and carbon nanofibers. *Macromolecules.* 2008;41(19):7090–7.
11. McNally T, Pötschke P, Halley P, Murphy M, Martin D, Bell SEJ, et al. Polyethylene multiwalled carbon nanotube composites. *Polymer.* 2005;46(19):8222–32.
12. Bonnaud L, Murariu O, Basso NRD, Dubois P. High viscosity polyethylene-based electroconductive nanocomposites: Carbon nanotubes versus carbon nanofibres. *Polym Bull.* 2013;70(3):895–904.
13. Kalakonda P, Iannacchione GS, Daly M, Georgiev GY, Cabrera Y, Judith R, et al. Calorimetric study of nanocomposites of multiwalled carbon nanotubes and isotactic polypropylene polymer. *J Appl Polym Sci.* 2013;130(1):587–94.
14. El Achaby M, Qaiss A. Processing and properties of polyethylene reinforced by graphene nanosheets and carbon nanotubes. *Mater Des.* 2013;44:81–9.
15. Chrissafis K, Paraskevopoulos KM, Tsiaoussis I, Bikiaris D. Comparative study of the effect of different nanoparticles on the mechanical properties, permeability, and thermal degradation mechanism of HDPE. *J Appl Polym Sci.* 2009;114(3):1606–18.
16. Wang Q, Dai J, Li W, Wei Z, Jiang J. The effects of CNT alignment on electrical conductivity and mechanical properties of SWNT/epoxy nanocomposites. *Compos Sci Technol.* 2008;68(7):1644–48.
17. Nanni F, Mayoral BL, Madau F, Montesperelli G, McNally T. Effect of MWCNT alignment on mechanical and self-monitoring properties of extruded PET–MWCNT nanocomposites. *Compos Sci Technol.* 2012;72(10):1140–6.
18. Zhang R, Dowden A, Deng H, Baxendale M, Peijs T. Conductive network formation in the melt of carbon nanotube/thermoplastic polyurethane composite. *Compos Sci Technol.* 2009;69(10):1499–504.
19. Xiang D, Guo JD, Kumar A, Chen BQ, Harkin-Jones E. Effect of processing conditions on the structure, electrical and mechanical properties of melt mixed high density polyethylene/multi-walled CNT composites in compression molding. *Mater Test.* 2017;59(2):136–47.

20. Du J, Zhao L, Zeng Y, Zhang L, Li F, Liu P, et al. Comparison of electrical properties between multi-walled carbon nanotube and graphene nanosheet/high density polyethylene composites with a segregated network structure. *Carbon.* 2011;49(4):1094–1100.
21. Verge P, Benali S, Bonnaud L, Minoia A, Mainil M, Lazzaroni R, et al. Unpredictable dispersion states of MWNTs in HDPE: A comparative and comprehensive study. *Eur Polym J.* 2012;48(4):677–83.
22. Xi Y, Yamanaka A, Bin Y, Matsuo M. Electrical properties of segregated ultrahigh molecular weight polyethylene/multiwalled carbon nanotube composites. *J Appl Polym Sci.* 2007;105(5):2868–76.
23. Tao FF, Bonnaud L, Auhl D, Struth B, Dubois P, Bailly C. Influence of shear-induced crystallization on the electrical conductivity of high density polyethylene carbon nanotube nanocomposites. *Polymer.* 2012;53(25):5909–16.
24. Chatterjee S, Nuesch FA, Chu BTT. Comparing carbon nanotubes and graphene nanoplatelets as reinforcements in polyamide 12 composites. *Nanotechnology.* 2011;22(27):275714.
25. Kalaitzidou K, Fukushima H, Drzal LT. A new compounding method for exfoliated graphite–polypropylene nanocomposites with enhanced flexural properties and lower percolation threshold. *Compos Sci Technol.* 2007;67(10):2045–51.
26. Yu J, Zhang LQ, Rogunova M, Summers J, Hiltner A, Baer E. Conductivity of polyolefins filled with high-structure carbon black. *J Appl Polym Sci.* 2005;98(4):1799–805.
27. Wen M, Sun X, Su L, Shen J, Li J, Guo S. The electrical conductivity of carbon nanotube/carbon black/polypropylene composites prepared through multistage stretching extrusion. *Polymer.* 2012;53(7):1602–10.
28. Kalaitzidou K, Fukushima H, Askeland P, Drzal LT. The nucleating effect of exfoliated graphite nanoplatelets and their influence on the crystal structure and electrical conductivity of polypropylene nanocomposites. *J Mater Sci.* 2008;43(8):2895–907.
29. Chen B, Evans JRG. Nominal and effective volume fractions in polymer–clay nanocomposites. *Macromolecules.* 2006;39(5):1790–6.
30. Haznedar G, Cravanzola S, Zanetti M, Scarano D, Zecchina A, Cesano F. Graphite nanoplatelets and carbon nanotubes based polyethylene composites: Electrical conductivity and morphology. *Mater Chem Phys.* 2013;143(1):47–52.
31. Kim BJ, Byun JH, Park SJ. Effects of graphenes/CNTs co-reinforcement on electrical and mechanical properties of HDPE matrix nanocomposites. *Bull Korean Chem Soc.* 2010;31:2261–4.
32. Das D, Satapathy BK. Microstructure-rheological percolation-mechanical properties correlation of melt-processed polypropylene-multiwall carbon nanotube nanocomposites: Influence of matrix tacticity combination. *Mater Chem Phys.* 2014;147(1–2):127–40.
33. Du F, Scogna R, Zhou W, Brand S, Fischer J, Winey K. Nanotube networks in polymer nanocomposites: rheology and electrical conductivity. *Macromolecules.* 2004;37:9048–55.
34. Du F. Nanotube networks in polymer nanocomposites: Electrical and thermal properties, rheology, and flammability. Ph.D., University of Pennsylvania, Ann Arbor, 2005.

3 Biaxially Stretched Conductive Polymer Composites

3.1 INTRODUCTION

In general, the addition of nanofillers can effectively improve the mechanical, thermal, electrical, and barrier properties of polymers. In addition, the preparation process and processing parameters of composites also have a great impact on the structure and properties of composites. Biaxial stretching film technology has the advantages of high production efficiency and stable quality, which makes the polymer chain alignment along the stretching direction, so as to adjust and improve the aggregation structure of polymer. It has advantages in strength, rigidity, stability, optical properties, thickness uniformity, and so on. It has become the most important advanced green technology in film manufacturing.

Biaxial stretching is not only an advanced film manufacturing process but also a deformation mode in some processing methods, such as blow molding, thermoforming, and extrusion [1–4]. The deformation is essentially biaxial elongation in a semi-solid or molten state, and it is an easy and efficient method to enhance the properties of polymer composites. Biaxial stretching used in changing structure, dispersion, and orientation of neat polymer materials to enhance performances has been widely studied, such as biaxially oriented polypropylene (BOPP) [5], biaxially oriented polyethylene (BOPE) [6], and biaxially oriented polyethylene terephthalate (BOPET) [7]. However, few studies [1,4,8,9] investigate the effect of the process of biaxial stretching on the performance improvement of composites and applications of biaxially stretched polymer/carbon nanofiller nanocomposites.

In this chapter, the effect of biaxial stretching process on structure change and final performances of high-density polyethylene (HDPE)/carbon nanofiller composites are investigated and discussed. First, the structure evolution, tensile, electrical, and thermal properties of HDPE/multi-walled carbon nanotube (MWCNT) nanocomposite at different stretching modes (sequential (seq-) and simultaneous (sim-) biaxial stretching), stretching ratios (SRs), strain rates (sr), and temperatures (T) are studied and analyzed. Then, a comparative study on the deformation behavior, structural evolution, and properties of biaxially stretched HDPE/carbon nanofillers composites with 4 wt% MWCNTs, graphene nanoplatelets (GNPs), or carbon blank (CB) was carried out to investigate the influence of the carbon nanofillers with different dimensions on the processability of the material and on its final properties after biaxial stretching. Furthermore, biaxially stretched binary carbon nanofiller-reinforced HDPE nanocomposites were conducted at different SRs to explore the synergistic

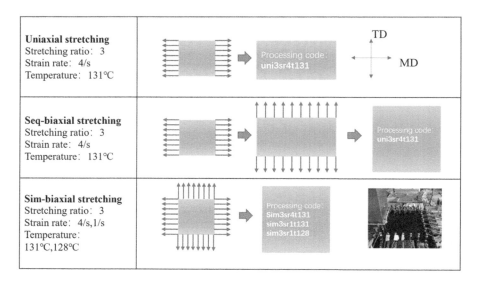

FIGURE 3.1 Schematic of uniaxial, seq-biaxial, and sim-biaxial stretching with related processing parameters (inset: a photograph of the sim-biaxial stretching process of HDPE/MWCNT composite sheet). HDPE, high-density polyethylene; MWCNT, multi-walled carbon nanotube.

effects between binary carbon nanofillers after tension and the influence of biaxial stretching process on the structure and properties of the HDPE/carbon nanofiller composites. The processes of uniaxial, seq-biaxial, and sim-biaxial stretching modes with different conditions are shown in Figure 3.1 [10].

3.2 BIAXIALLY STRETCHED HDPE/MWCNT COMPOSITES

First, the influence of the addition of MWCNTs on the biaxial deformation behavior of sheet samples was studied. Furthermore, the structure evolution and properties of biaxially stretched HDPE/MWCNT composites and the effect of different stretching parameters on the properties were investigated. In addition, some uniaxial stretching of the HDPE/MWCNT composite with 4 wt% MWCNTs was also conducted in order to better understand the effect of nanotube alignment on the properties of the composites.

3.2.1 Biaxial Deformation Behavior

Figure 3.2 shows the uniaxial, seq-biaxial, and sim-biaxial strain-stress curves for stretched neat HDPE sheets and composite sheets containing 4 wt% MWCNTs at an SR of 3. Figure 3.3 shows the data in bar chart form and includes the initial sheet modulus and strain hardening index of the material during uniaxial, seq-biaxial, and sim-biaxial stretching.

A large increase in stretching stress and a significant strain hardening can be observed during the uniaxial stretching of composite sheets with 4 wt% MWCNTs

Biaxially Stretched Conductive Polymer Composites

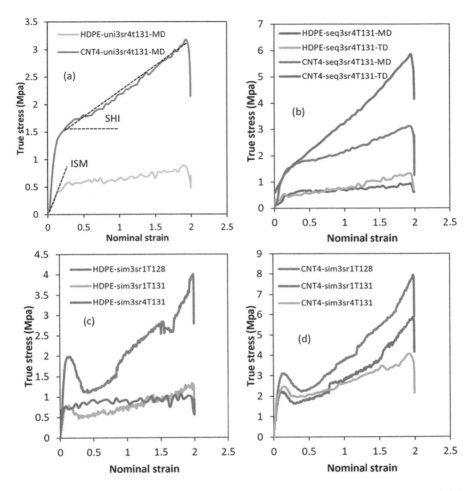

FIGURE 3.2 Strain-stress curves of (a) composite and neat HDPE sheet during unixial stretching; (b) seq-biaxial stretching, and (c and d) sim-biaxial stretching (MD and TD refer to the machine direction and transverse direction, respectively). HDPE, high-density polyethylene; MD, machine direction; TD, transverse direction.

in Figure 3.2a. According to Figure 3.3, the initial sheet modulus (ISM) and strain hardening index (SHI) of the material increased by 285% and 293%, respectively, during uniaxial stretching with the addition of 4wt% MWCNTs. The sequential biaxial stretching process can be regarded as a double continuous uniaxial stretching process; hence, one can see that the strain-stress curves (Figure 3.2b) and the stretching properties in the machine direction (MD) (Figure 3.3) during the seq-biaxial stretching of the composite and neat HDPE sheets are very similar to the ones during uniaxial stretching. However, the deformation behavior in the transverse direction (TD) is very different to that in the MD, especially for the composite sheet. There is almost no elastic deformation or yield behavior in the TD after the first stretching

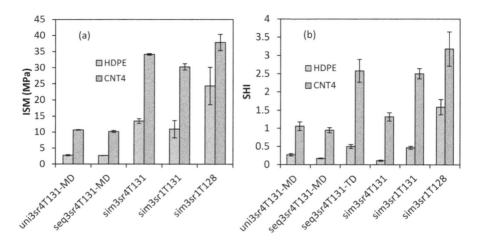

FIGURE 3.3 Influence of MWCNTs on the (a) initial sheet modulus and (b) strain hardening index of material during uniaxial, seq-biaxial, and sim-biaxial stretching. MWCNTs, multi-walled carbon nanotubes.

process in the MD, and a greater strain hardening effect can be seen in Figure 3.2b. The SHI of the composite in the TD increased by 388% compared with the neat HDPE according to Figure 3.3b.

The strain-stress curves for sim-biaxial stretching with different stretching conditions are shown in Figure 3.2c and d, respectively. As stated earlier, the strain hardening is very weak for the neat HDPE at a stretching temperature of 131°C and a strain rate of $4\,s^{-1}$. In Figure 3.2c, it can be observed that an yield peak and some strain hardening occurs when the strain rate decreases to $1\,s^{-1}$, which may be attributed to a slower but more stable disentanglement process of the HDPE chains. The yield peak and strain hardening become much more obvious for the neat HDPE sheet at a stretching temperature of 128°C and a strain rate of $1\,s^{-1}$, indicating that the residual crystallites (21.3%) have a significant influence on the deformation behavior of the neat HDPE sheet. In Figure 3.2d, similar phenomena for the composite with 4 wt% MWCNTs can also be observed. The average values of ISM and SHI increase by 190% and 462%, respectively, during the uniaxial, seq-biaxial, and sim-biaxial stretching of the composite sheet.

3.2.2 Structural Evolution

The morphology of stretched HDPE/MWCNT composites was examined using SEM after plasma etching the samples [4]. Figure 3.4 shows the SEM images of composites containing 8 wt% MWCNTs sim-biaxially stretched at a strain rate of $4\,s^{-1}$ and the temperature of 131°C with increasing SRs. The existence of numerous MWCNT network structures consisting of agglomerated and isolated nanotubes in the HDPE matrix before stretching can be observed in Figure 3.4a. Fewer MWCNT agglomerates are observed after biaxial stretching at an SR of 2 (Figure 3.4b), because some individual nanotubes are pulled out of the agglomerates upon stretching.

Biaxially Stretched Conductive Polymer Composites 47

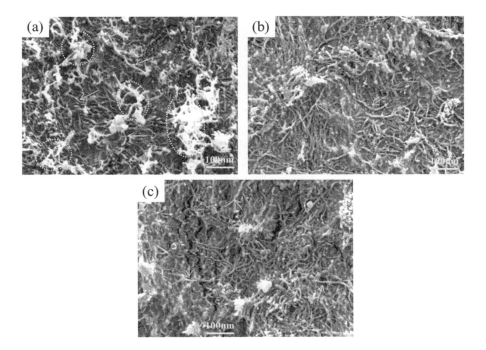

FIGURE 3.4 SEM images of HDPE/MWCNT composites containing 8 wt% MWCNTs at (a) SR = 1 (unstretched), (b) SR = 2, and (c) SR = 3 (the agglomerated nanotubes are circled, and the individual nanotubes are indicated by arrows in the micrographs). HDPE, high-density polyethylene; MWCNT, multi-walled carbon nanotube; SRs, stretching ratios.

The MWCNTs appear to be randomly oriented in the stretching plane after disentanglement. A further breakup of MWCNT agglomerates can be observed as the SR increases to 3 in Figure 3.4c, and the MWCNTs are further oriented in the stretching plane. However, the composite containing 8 wt% MWCNTs still has many MWCNT networks remaining at an SR of 3. The three-dimensional MWCNT networks tend to transform to two-dimensional network with increasing SR.

Figure 3.5 shows the SEM images of HDPE/MWCNT composites containing 4 wt% MWCNTs stretched at different stretching parameters [10]. From Figure 3.5a, there are some agglomerated and isolated MWCNTs in the unstretched sample. It can be clearly observed in Figure 3.5b that the MWCNT agglomerates and individual MWCNTs are well oriented along the uniaxial stretching direction. As shown in Figure 3.5c, the morphology of the seq-biaxially stretched composite in which fewer MWCNT agglomerates can be observed is compared with the uniaxially stretched composite. This is due to the greater disentanglement potential of stretching in two directions. The MWCNTs appear randomly oriented in the stretching plane after seq-biaxially stretching. Figure 3.5d–f shows the sim-biaxially stretched composite sheet with 4 wt% MWCNTs at sim3sr4T131, sim3sr1T131, and sim3sr1T128, respectively. It can be seen from Figure 3.5d–f that the orientation of the nanotubes is also random after sim-biaxially stretching.

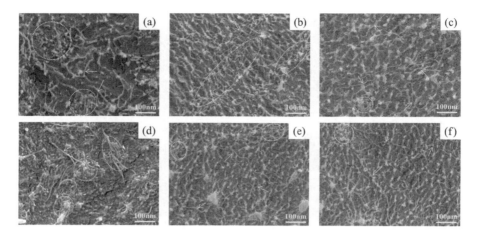

FIGURE 3.5 SEM images of HDPE/MWCNT composites containing 4 wt% MWCNTs with different stretching parameters: (a) unstretched, (b) uni3sr4T131, (c) seq3sr4T131, (d) sim3sr4T131, (e) sim3sr1T131, and (f) sim3sr1T128. HDPE, high-density polyethylene; MWCNT, multi-walled carbon nanotube.

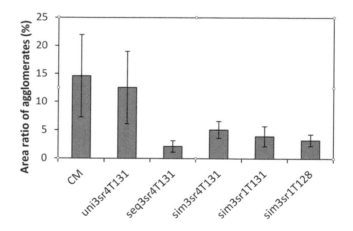

FIGURE 3.6 Area ratio of agglomerates for the HDPE/MWCNT composites as measured from SEM results. HDPE, high-density polyethylene; MWCNT, multi-walled carbon nanotube.

In order to quantitatively analyze the disentanglement of MWCNT agglomerates during uniaxial, seq-biaxial, and sim-biaxial stretching, the area ratio of agglomerates (A'_R) was calculated from the SEM results in JMicroVision digital image processing software. In this calculation, the disentangled agglomerates were regarded as individual nanotubes, and the results are shown in Figure 3.6.

Here it can be seen that the area ratio of agglomerates only slightly decreases by about 2% for the uniaxially stretched composite sheet due to insufficient deformation to break the nanotube agglomerates. However, a significant decrease of about 12% in

Biaxially Stretched Conductive Polymer Composites

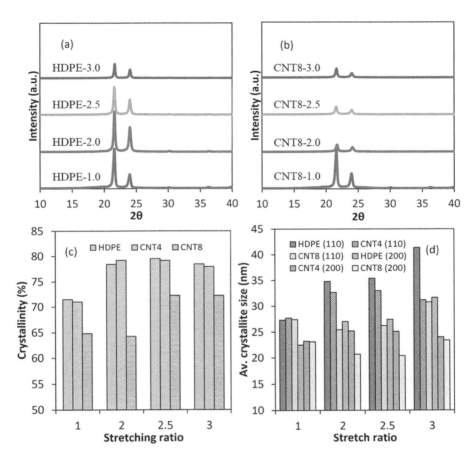

FIGURE 3.7 (a and b) WAXRD spectra, (c) crystallinity, and (d) average crystallite sizes of HDPE and HDPE/MWCNT composite sheets with increasing SRs. HDPE, high-density polyethylene; MWCNT, multi-walled carbon nanotube.

area ratio of agglomerates can be seen for the seq-biaxially stretched composite sheet indicating that many individual nanotubes are pulled out of the agglomerates upon seq-biaxial stretching. A clear decrease in the area ratio of agglomerates can also be seen for all the sim-biaxially stretched samples, and more MWCNT agglomerates are disentangled at the strain rate of $1\ s^{-1}$ and stretching temperature of 128°C due to the higher stresses transferred at these conditions.

Figure 3.7a and b shows the wide angle X-ray diffraction (WAXRD) spectra for the unstretched and stretched HDPE and composites with 8 wt% MWCNTs at different SRs. It can be observed that the two intensive reflection peaks become less intensive after biaxial deformation in Figure 3.7a and b due to the interference of numerous imperfect crystallites, especially for the samples with 8 wt% MWCNTs as a result of the restriction effect of MWCNT network on the mobility of polymer chains [11]. The two weak peaks are almost undetectable for both filled and unfilled HDPE after biaxial stretching.

In general, the crystallinity of all the biaxially stretched samples increases by about 8% due to strain-induced crystallization, as shown in Figure 3.7c. The lower crystallinities of composites with 8 wt% MWCNTs compared to other samples at all the SRs indicate that the addition of abundant MWCNTs has an inhibitive effect on the growth of crystallites. The results of crystallinity from WAXRD agree well with those from the differential scanning calorimetry (DSC) tests in Section 3.2.3. Figure 3.7d shows the average crystallite sizes in the (110) and (200) reflection planes from the Scherrer Equation $\left(L_{hkl} = \dfrac{K\lambda}{\beta\cos\theta} \right)$ [12] for the stretched HDPE and composites at increasing SRs. The results show an increasing trend in the average crystallite sizes for all the stretched materials with increasing SR, but the composites with

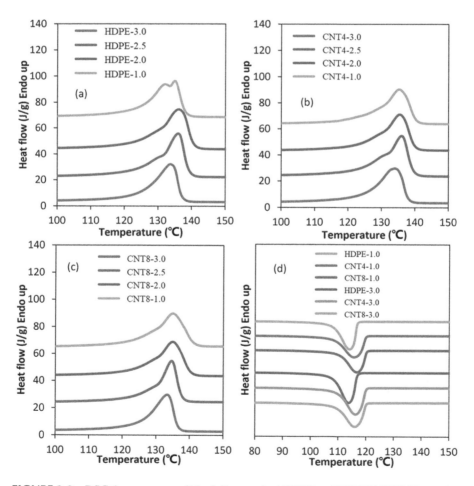

FIGURE 3.8 DSC thermograms of biaxially stretched HDPE and HDPE/MWCNT samples with different SRs at (a–c) the first heating and (d) cooling stages. HDPE, high-density polyethylene; MWCNT, multi-walled carbon nanotube; SRs, stretching ratios.

8 wt% MWCNTs have a lesser increase. The presence of imperfect crystallites and the increase in average crystallite sizes imply a broadened size distribution of crystallites in the stretched samples. This is also supported by the following DSC results.

3.2.3 Thermal Properties

The melting and crystallization behaviors of biaxially stretched samples were investigated by DSC. Some additional structural information can be revealed from the thermal behavior of the deformed samples in the heating and crystallization stages, particularly, the first heating stages. The first heating and cooling curves and relevant thermal parameters for the stretched and unstretched samples are shown in Figure 3.8 and Table 3.1, respectively. A shoulder on the low temperature side of the melting peak for the stretched samples can be observed in Figure 3.8a–c. This phenomenon can be explained by the less perfect crystallites with a smaller size melting at lower temperatures [8,9]. Figure 3.8d shows some typical DSC cooling curves of stretched and unstretched samples. It can be observed that the crystallization behavior is not significantly influenced by biaxial stretching.

As shown in Table 3.1, the crystallinity (X_c) of stretched neat HDPE and HDPE/MWCNT composite with 4 wt% MWCNTs obtained from the first heating process increases by about 10%–15% as a result of strain-induced crystallization, while the crystallinity of stretched composites with 8 wt% MWCNTs increases to a lesser degree (by ~10%) due to the confinement effect of high MWCNT loading on the growth of crystallites. The slightly higher crystallinity from DSC compared with that from WAXRD may be due to the recrystallization of some imperfect crystallites during the DSC heating process. A slight increase (about 1°C–2°C) in melting temperature (T_m) can be found

TABLE 3.1
Thermal Parameters of the Unstretched and Biaxially Stretched Samples Obtained from DSC Tests

Sample	SR	X_c^{1st} (%)	X_c^{2st} (%)	T_m^{1st} (°C)	W_h^{1st} (°C)	T_c (°C)
HDPE	1.0	72.8 ± 3.1	70.4 ± 1.3	133.9 ± 0.2	7.4 ± 0.1	114.0 ± 0.2
	2.0	86.6 ± 0.6	77.8 ± 2.1	136.4 ± 0.3	7.6 ± 0.2	113.1 ± 0.3
	2.5	86.0 ± 2.9	76.5 ± 2.3	135.7 ± 0.4	8.2 ± 0.4	113.7 ± 0.4
	3.0	85.7 ± 2.9	76.3 ± 1.4	135.2 ± 0.2	9.2 ± 0.2	113.6 ± 0.3
CNT4	1.0	73.9 ± 5.1	70.7 ± 2.4	134.0 ± 0.1	7.7 ± 0.3	116.0 ± 0.2
	2.0	84.4 ± 5.0	79.6 ± 1.8	135.9 ± 0.4	8.0 ± 0.1	115.7 ± 0.4
	2.5	86.1 ± 2.1	78.5 ± 1.1	136.1 ± 0.7	8.1 ± 0.1	115.7 ± 0.5
	3.0	85.5 ± 3.0	78.2 ± 6.1	135.4 ± 0.2	9.0 ± 0.4	115.9 ± 0.4
CNT8	1.0	70.9 ± 3.5	72.6 ± 1.9	133.6 ± 0.3	7.1 ± 0.3	116.2 ± 0.4
	2.0	78.0 ± 1.3	76.6 ± 1.3	134.7 ± 0.5	5.5 ± 0.7	116.5 ± 0.3
	2.5	80.7 ± 3.4	77.3 ± 2.1	135.1 ± 0.3	8.9 ± 0.4	115.8 ± 0.3
	3.0	81.5 ± 3.6	77.5 ± 6.7	134.8 ± 0.2	9.7 ± 0.5	115.9 ± 0.1

CNT, carbon nanotube; HDPE, high-density polyethylene.

in the first heating stage as a result of the thickening of lamella generated by the oriented polymer chains [1,13]. The increased width at half height (W_h) in the first heating curves of stretched samples indicates the wider distribution of crystallite sizes [14]. Compared to neat HDPE, the crystallization temperature (T_c) of HDPE/MWCNT composites, regardless of being stretched or unstretched, increases by around 2°C.

3.2.4 Tensile Properties

The mechanical properties of biaxially stretched sheets were investigated by tensile testing, as shown in Figure 3.9 and Table 3.2. It can be seen in Figure 3.9a–c that the experimental values of Young's modulus (E), stress at yield (σ_y), and stress at break (σ_b) of stretched HDPE/MWCNT composites increase steadily with increasing SRs due to a combination of MWCNT agglomerate disentanglement, polymer molecular orientation, and MWCNT orientation. The modulus, σ_b, and ε_b of the composite with 4 wt% MWCNTs stretched at sim3sr4T131 increased 27%, 228%, and 46%, respectively. E, σ_y, and σ_b also increase for the pure HDPE sheets up to an SR of 2.5 after which they drop off. This drop is likely to be due to the relaxation of polymer chains prior to solidification which is not observed in the materials containing MWCNTs due to restricted molecular mobility imposed by the oriented nanotubes. The slight temperature increase caused by adiabatic heating is unlikely to significantly affect the relaxation process.

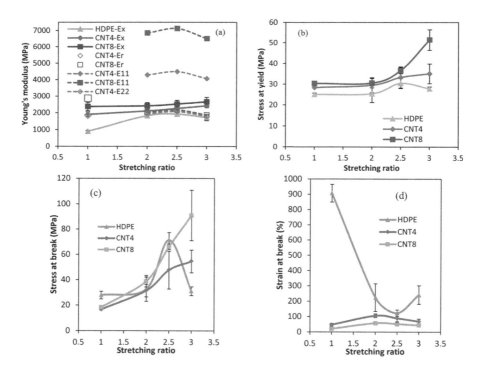

FIGURE 3.9 Effect of biaxial stretching on (a) Young's modulus, (b) stress at yield, (c) stress at break, and (d) strain at break of the different carbon nanofiller-reinforced composites with increasing SRs. SRs, stretching ratios.

TABLE 3.2
Changes in the Experimental Values of Tensile Properties of Biaxially Stretched Samples with the Addition of MWCNTs at a Strain Rate of 4 s⁻¹ and a Stretching Temperature of 131°C

Sample	SR	ΔE (%)	$\Delta\sigma_y$ (%)	$\Delta\sigma_b$ (%)	$\Delta\varepsilon_b$ (%)
CNT4-2.0	2.0	+15.4	+16.6	−3.4	−52.6
CNT8-2.0	2.0	+32.1	+20.6	+19.3	−74.4
CNT4-2.5	2.5	+17.6	+9.6	−33.2	−27.0
CNT8-2.5	2.5	+31.1	+20.6	−7.3	−57.7
CNT4-3.0	3.0	+40.0	+26.5	+75.5	−71.6
CNT8-3.0	3.0	+53.8	+85.4	+192.7	−81.8

CNT, carbon nanotube; HDPE, high-density polyethylene; MWCNT, multi-walled carbon nanotube.

The experimental modulus values of stretched composites are also compared with the predicted values from the randomly aligned Halpin–Tsai model $\left(\dfrac{E_r}{E_m} = \dfrac{3}{8}\left[\dfrac{1+\xi\eta_L\varphi_f}{1-\eta_L\varphi_f}\right] + \dfrac{5}{8}\left[\dfrac{1+2\eta_T\varphi_f}{1-\eta_T\varphi_f}\right]\right)$. The results are also shown in Figure 3.9a. The predicted values are much higher than the experimental values for the stretched composite containing 8 wt% MWCNTs. This difference can be attributed to the fact that aggregated nanotubes still exist after deformation, thus resulting in a lower than predicted reinforcing effect. The errors between the predicted and experimental modulus values at an SR of 3 are smaller for the composite with 4 wt% MWCNTs, since agglomeration is expected to be less for the lower loading.

The ε_b of stretched composites increases by about 100% (Figure 3.9d), which is mainly attributed to the breakup of MWCNT agglomerates and removal of stress concentrators. However, it decreases by ~90% for the stretched HDPE at an SR = 2.5 due to the orientation of polymer chains and reduced potential for further stretching. The increase in the ε_b of stretched HDPE at an SR of 3 can be attributed to the relaxation of polymer chains. Overall, a higher SR has a very positive influence on the mechanical properties of the HDPE/MWCNT composites. For instance, for the stretched composite with 8 wt% MWCNTs at an SR of 3, the E and σ_b increase by about 54% and 193%, respectively, compared with the unfilled HDPE at the same SR (Table 3.3). Therefore, the disentanglement and orientation of MWCNTs, rather than the orientation of polymer chains, dominates the changes in the mechanical properties of stretched HDPE/MWCNT composites.

3.2.5 ELECTRICAL PROPERTIES

The variations in the volume resistivity of biaxially stretched HDPE/MWCNT composites with different MWCNT loadings as a function of SR were investigated, as shown in Figure 3.10a. And the shifts in percolation threshold of the stretched HDPE/MWCNT

TABLE 3.3
The Critical Concentration ϕ_t and Critical Exponent t of Biaxially Stretched HDPE/MWCNT Composites at Different SRs

Parameter	SR = 1.0	SR = 2.0	SR = 2.5	SR = 3.0
ϕ_t	1.9	3.8	3.8	4.9
t	1.9	1.6	1.5	1.1

HDPE, high-density polyethylene; MWCNT, multi-walled carbon nanotube; SRs, stretching ratios.

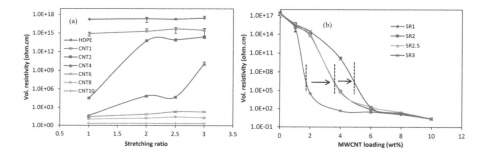

FIGURE 3.10 Variations in the volume resistivity of biaxially stretched HDPE/MWCNT composites as a function of (a) SR and (b) MWCNT loading. HDPE, high-density polyethylene; MWCNT, multi-walled carbon nanotube; SRs, stretching ratios.

composites at different SRs are plotted in Figure 3.10b. As shown in Figure 3.10a, the resistivity of neat HDPE and the composite with 1 wt% MWCNTs is barely influenced by biaxial stretching. However, the resistivity of the composite with 2 wt% MWCNTs increased by nine orders of magnitude at an SR of 2 before leveling off. Presumably, at this SR, the distance between the nanotubes exceeds the critical maximum distance (about 1.8 nm [15]) for electron hopping. For the composite with 4 wt% MWCNTs, the resistivity does not increase significantly until an SR of 2.5 is exceeded when it increases rapidly by about six orders of magnitude when the SR increases to 3.

Shen et al. [1,9] also observed that the resistivity of PP/CNT composites increased upon initial stretching but as the SR increased beyond 2.5, the resistivity began to decrease again due to the rebuilding of nanotube conductive networks. This rebuilding process of nanotube networks is not found in the current work about biaxial stretching HDPE-based composites, which indicates that the polymer matrix may also have a significant influence on the conductive networks. For example, the high viscosity of the HDPE used in this study is likely to inhibit the bridging of disentangled nanotubes. It may also be that the rapid crystallization of HDPE prevents relaxation of molecular orientation at the end of the stretching process. A slowly crystallizing polymer such as PP would allow for some orientation relaxation and thus greater potential for CNT bridging. There are no obvious variations in the resistivity

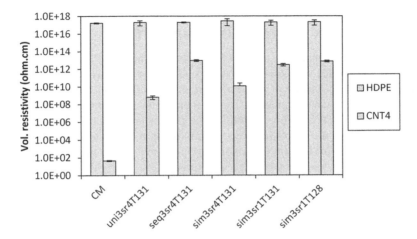

FIGURE 3.11 Changes in the resistivity of stretched HDPE/MWCNT samples at different stretching parameters. HDPE, high-density polyethylene; MWCNT, multi-walled carbon nanotube.

of composite containing MWCNT loadings higher than 6 wt% after biaxial stretching at increasing SRs. One might assume that at these loadings there is a sufficient density of nanotubes to maintain a critical distance between nanotubes regardless of the strain thus forming a robust conductive network in the matrix.

According to the scaling law of percolation threshold, the unstretched HDPE/MWCNT composites exhibit a low critical concentration of 1.9 wt% MWCNTs and a typical three-dimensional system ($t = 1.9$). The electrical resistivity of stretched samples was also fitted according to the scaling law. The values of ϕ_t and t from fitting are shown in Table 3.3. The gradually increasing values of ϕ_t imply the destruction of conductive networks with increasing SRs, and the decreased values of t reveal a transformation of system structure from three-dimensional to two-dimensional.

The volume resistivity of samples stretched at different stretching parameters was measured in the thickness direction using a high resistivity electrometer, and the results are shown in Figure 3.11.

The resistivity of all the neat HDPE samples stretched at different stretching parameters is not changed. However, these stretching parameters influence the resistivity of the stretched composite with 4 wt% MWCNTs. The correlation between the area ratio of agglomerate A'_R (Figure 3.6) and log resistivity of the stretched composite samples is −0.93 with a p value of 0.01. This indicates that fewer secondary MWCNT agglomerates will lead to higher resistivity. In Socher's [16], Huang's [17], and Alig's [18] research, it has been reported that secondary nanotube agglomerates can enhance the conductive networks. The resistivity of uniaxially and sim-biaxially stretched composites stretched at a strain rate of 4 s[−1] and stretching temperature of 131°C increases by ~10^7 and ~10^9 Ω·cm, respectively. However, due to more disentanglement of secondary nanotube agglomerates, the resistivity of the seq-biaxially stretched composite samples and the sim-biaxially stretched composite samples stretched at a strain rate of 1 s[−1] increases by ~10^{12} and ~10^{11} Ω·cm, respectively.

3.3 BIAXIALLY STRETCHED UNARY CARBON NANOFILLER-REINFORCED COMPOSITES

In this section, a comparative study of biaxially stretched HDPE/carbon nanofiller-reinforced composites with 4 wt% MWCNTs, GNPs, or CB was performed in order to investigate the influence of different carbon nanofillers on the performance of biaxially stretched nanocomposites. The final mechanical, electrical, and gas barrier properties of the HDPE/carbon nanofiller composites are affected by biaxial deformation and the magnitude of the effect depending on the dimensional shapes of the carbon nanofillers. In this study, the sim-biaxially stretched samples were stretched at a stretching temperature of 131°C; a strain rate of $4\,s^{-1}$; and SRs of 2, 2.5, and 3.

3.3.1 Biaxial Deformation Behavior

The deformation behavior of carbon nanofillers with different dimensions reinforced composites during the process of biaxial stretching is compared, as shown in Figure 3.12. A remarkable strain hardening behavior for the 4 wt% carbon nanofiller composites upon biaxial stretching can be observed in Figure 3.12a, and the reinforcement effectiveness of the nanofillers in strain hardening behavior is CB < MWCNTs < GNPs.

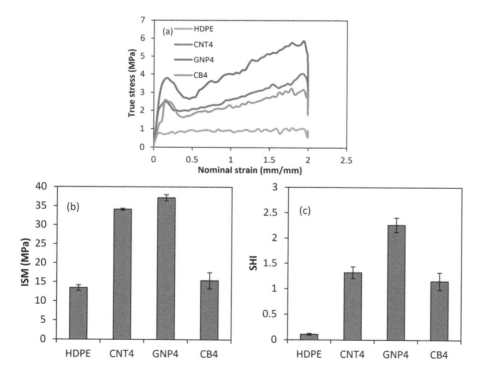

FIGURE 3.12 Effect of carbon nanofillers with different dimensions on the (a) strain-stress curves, (b) ISM, and (c) SHI of the unary carbon nanofiller-reinforced composites during biaxial stretching. ISM, initial sheet modulus; SHI, strain hardening index.

The tendency of the ISM and SHI of the HDPE/carbon nanofiller composites during biaxial stretching is homologous, as shown in Figure 3.12b and c, respectively.

3.3.2 THERMAL PROPERTIES

The melting and crystallization behavior of the biaxially stretched unary carbon nanofiller-reinforced composites was investigated by DSC [19]. The relevant thermal parameters obtained from the first heating and cooling stages are shown in Figure 3.13. Regardless of the types of carbon nanofiller, the crystallinity of all the stretched samples obtained from the first heating process increases by about 10%–15% (Figure 3.13a), which is slightly higher than the crystallinity from WAXRD.

In Figure 3.13b, a slight increase (about 1°C–2°C) in T_m can be observed in the first heating stage for all the stretched samples due to the presence of crystallites with increased size. The smaller W_h for the stretched HDPE/GNP and HDPE/CB composites indicates that there might be a narrower distribution of crystallite sizes in these systems compared to the unfilled and 4 wt% MWCNT-filled HDPE (Figure 3.13c).

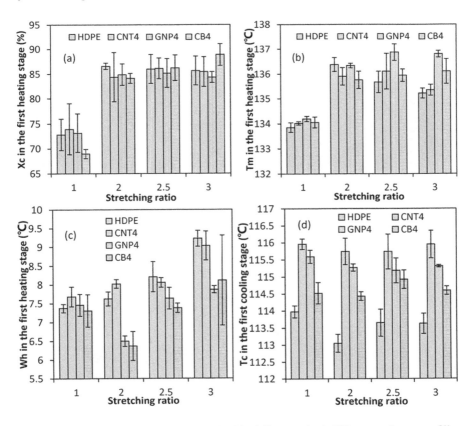

FIGURE 3.13 Thermal parameters of the biaxially stretched different carbon nanofiller-reinforced composites with increasing SRs: (a) crystallinity, (b), melting temperature, (c) width at half height, (d) crystallization temperature. SRs, stretching ratios

It can be observed in Figure 3.13d that the crystallization temperatures of the different carbon nanofiller-reinforced composites are not significantly affected by biaxial deformation.

3.3.3 STRUCTURAL EVOLUTION

The morphology of unstretched and biaxially stretched HDPE/carbon nanofiller composites containing 4 wt% MWCNTs was observed by SEM, as shown in Figure 3.5 in Section 3.2.2. And the SEM images of the unstretched and stretched HDPE/GNP and HDPE/CB composites are shown in Figure 3.14. It can be seen in Figure 3.14a and b that the GNPs and CBs are typical two-dimensional and zero-dimensional shape characteristics, respectively, which are to aggregate due to strong van der Waals forces. From Figure 3.14c and d, the randomly aligned GNPs in HDPE/GNP composite tend to align parallel to the surface of the sample with much less wrinkles after biaxial stretching, and the exfoliation or slippage of GNPs could occur to some extent under the shear force. The clustered CB in HDPE/CB composite is dispersed more evenly in the matrix with increasing SRs.

The changes in the crystalline structure and percentage crystallinity of neat HDPE and HDPE/carbon nanofiller composites before and after biaxial stretching were investigated by WAXD [19]. Figure 3.15a–d shows the WAXD spectra for the unstretched

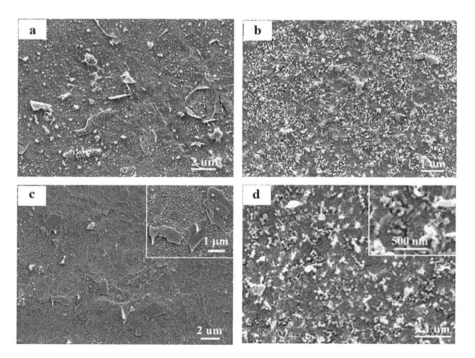

FIGURE 3.14 SEM images of unstretched (a) HDPE/GNP, (b) HDPE/CB composites and stretched (c) HDPE/GNP, and (d) HDPE/CB composites (SR = 3). CB, carbon blank; GNP, graphene nanoplatelet; HDPE, high-density polyethylene; SRs, stretching ratios.

Biaxially Stretched Conductive Polymer Composites

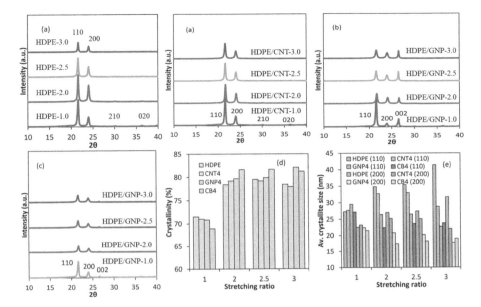

FIGURE 3.15 (a–d) WAXRD spectra, (e) crystallinities, and (f) average crystallite sizes of neat HDPE and HDPE/carbon nanofiller composite sheets with increasing SRs.

and stretched HDPE and composites containing 4 wt% carbon nanofillers at different SRs. Figure 3.15e and f shows the changes in crystallinities and average crystallite sizes of the samples after biaxially stretching with increasing SRs, respectively. From Figure 3.15a–d, the intensity of the reflection peaks of stretched neat HDPE becomes less intensive due to the interference of numerous imperfect crystallites. In addition, the reflection peaks of the stretched HDPE/GNP and HDPE/CB composites are obviously weaker compared to unstretched samples probably due to a stronger restriction effect of GNPs and CB on the mobility of polymer chains resulting in more imperfect crystallites. This is supported by the smaller average crystallite sizes for the stretched HDPE/GNP and HDPE/CB composites compared to the HDPE/MWCNT composite in Figure 3.15f. As shown in Figure 3.15e, the crystallinity of the unstretched samples is barely changed with the addition of carbon nanofillers, but that of all the biaxially stretched samples increase by about 8%–12% regardless of the shape characteristics of the carbon nanofillers. This indicates that strain-induced crystallization dominates the increase in crystallinity for the stretched samples.

The melting and crystallization behavior of the biaxially stretched HDPE/carbon nanofiller composites were investigated by DSC [19]. The relevant thermal parameters obtained from the heating and cooling stages are shown in Figure 3.16. The crystallinity of all the biaxially stretched carbon nanofiller-reinforced composite samples obtained from the heating process increases by about 10%–15%, as shown in Figure 3.16a, which is slightly higher than the crystallinity from WAXD. In Figure 3.16b, a slight increase in T_m is observed in the heating stage for stretched samples due to the presence of crystallites with increased size. It can be observed

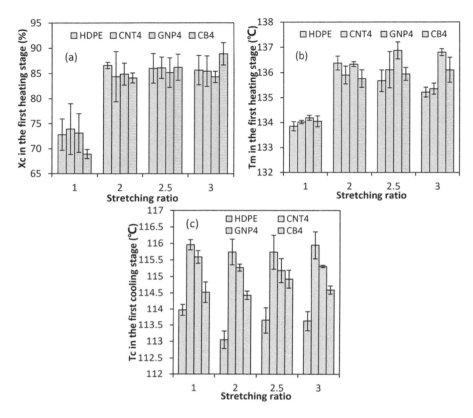

FIGURE 3.16 (a) Crystallinities, (b) melting temperature, and (c) crystallization temperature of the biaxially stretched HDPE/carbon nanofiller composites with increasing SRs. HDPE, high-density polyethylene; SRs, stretching ratios.

in Figure 3.16c that the T_c increased slightly for all the carbon nanofiller-reinforced composites compared with the neat HDPE due to the nucleation effect of carbon nanofiller. Besides, the T_c of neat HDPE and composites and the nucleation effect of nanofillers are not significantly affected by biaxial deformation. Furthermore, the higher T_c of the HDPE/MWCNT composite indicates that the one-dimensional nanotubes may more effectively facilitate the nucleation of HDPE.

3.3.4 Tensile Properties

The tensile properties of the unstretched and stretched HDPE/carbon nanofiller composites with the addition of 4 wt% carbon nanofillers were investigated, as shown in Figure 3.17 and Table 3.4. As shown in Figure 3.17a, the E of unstretched HDPE/nanofiller composites increased with the addition of 4 wt% carbon nanofillers. In Table 3.4, the unstretched composite filled with GNPs with the highest aspect ratio exhibits the highest reinforcement compared with the neat HDPE in modulus. Besides, the addition of carbon nanofillers has a negative effect on the σ_b and ε_b of

Biaxially Stretched Conductive Polymer Composites

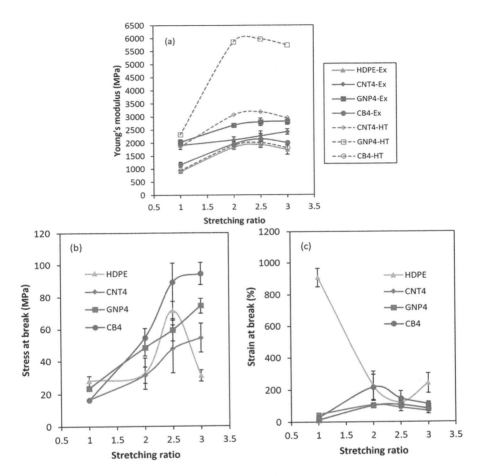

FIGURE 3.17 Effect of biaxial stretching on the (a) Young's modulus, (b) stress at break, and (c) strain at break of neat HDPE and HDPE/carbon nanofiller composites with increasing SRs. HDPE, high-density polyethylene; SRs, stretching ratios.

the unstretched sheets due to the presence of agglomerates, regardless of the dimensionality of the nanofiller.

In Figure 3.17a, the modulus of the stretched HDPE and HDPE/CB composite increases up to an SR of 2.5 followed by a decrease, indicating that the addition of CB does not dominate the changes in modulus for the stretched HDPE/CB composite samples due to the relatively weak reinforcement effectiveness of CB on modulus. On the contrary, due to a combination of the nanofiller deagglomeration and orientation which exert a restraining effect on the polymer [4,20], the modulus of HDPE/GNP and HDPE/MWCNT composites increases steadily with increasing SRs.

From Figure 3.17b and c, the σ_b and ε_b of all the stretched carbon nanofiller-reinforced composites are increased and stretched HDPE/CB composites exhibit the highest increase, which is probably attributable to a more intensive breakup of

TABLE 3.4
Effect of the Addition of 4 wt% Carbon Nanofillers on the Tensile Properties of Biaxially Stretched Samples at SRs of 2, 2.5, and 3

Sample	SR	ΔE (%)	$\Delta\sigma_b$ (%)	$\Delta\varepsilon_b$ (%)
CNT4	2.0	+15.4	−3.4	−52.6
GNP4		+46.0	+49.2	−54.3
CB4		+6.9	+68.7	−3.2
CNT4	2.5	+17.6	−33.2	−27.0
GNP4		+45.6	−16.6	−12.4
CB4		+11.4	+24.9	+20.3
CNT4	3.0	+40.0	+75.5	−71.6
GNP4		+63.1	+139.5	−65.6
CB4		+15.4	+204.4	−53.6

CB, carbon blank; CNT, carbon nanotube; GNP, graphene nanoplatelet; SRs, stretching ratios.

CB clusters during biaxial deformation. In addition, Table 3.4 shows that GNPs exhibit the most efficient reinforcement in E for the biaxially stretched samples, while the CB exhibits the most efficient reinforcement in σ_b. At an SR of 3, the E of the HDPE/GNP composite increased by about 63%, and the σ_b of the HDPE/CB composite increased by about 204%, compared with the neat HDPE at the same SR.

3.3.5 ELECTRICAL PROPERTIES

The volume resistivities of HDPE/carbon nanofiller composites with a concentration of 4 wt% as a function of SR during biaxial stretching are shown in Figure 3.18. The addition of carbon nanofillers efficiently decreases the resistivity of unstretched nanocomposites, and MWCNTs are efficient to generate conductive networks, which are likely to be due to one-dimensional particles easy to entangle and form networks. In addition, the resistivity of stretched HDPE/GNP composite slightly increases and levels off at a high resistivity can be observed in this study, which is different from the results obtained by Du et al. [21]. This is probably attributed to the distance between the GNPs being higher than the critical maximum distance for electron hopping after biaxial deformation. Furthermore, the resistivity of the stretched HDPE/CB composite increased by over 10^9 $\Omega\cdot$cm due to the destruction of the high structure of CB and increased interparticle distance. A similar increase in resistivity for unixially deformed PP/CB composites has been obtained by Deng et al. [22]. From the comparison of the resistivity of the biaxially stretched HDPE/carbon nanofiller composites with increasing SRs, finding that the MWCNT-filled composites exhibit a more robust conductive network as a result of more interlacing or entanglement of one-dimensional nanotubes.

Biaxially Stretched Conductive Polymer Composites 63

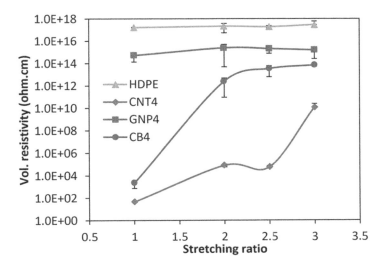

FIGURE 3.18 Volume resistivities of HDPE/carbon nanofiller composites with a concentration of 4 wt% as a function of SR during biaxial stretching. HDPE, high-density polyethylene; SR, stretching ratio.

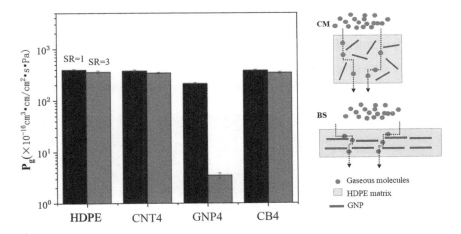

FIGURE 3.19 Permeability coefficient of the unstretched (a) and biaxially stretched samples (b) at SR = 3 and "Tortuous path" model for the HDPE/GNP composite. GNP, graphene nanoplatelet; HDPE, high-density polyethylene; SR, stretching ratio.

3.3.6 BARRIER PROPERTIES

Oxygen permeability of the unstretched and stretched samples at SR = 3 was characterized to investigate the influence of the addition of carbon nanofiller and biaxial stretching on the barrier properties of the materials, as shown in Figure 3.19. The permeability coefficient of unstretched HDPE/GNP composite drastically decreased by 64% due to the two-dimensional shape and very high aspect ratio of GNPs. The addition of

MWCNTs and CB can improve barrier properties of the polymer composites slightly, and biaxial stretching further improves the barrier properties of neat HDPE, HDPE/MWCNT, and HDPE/CB composites compared with unstretched samples, which is likely attributed to the increased crystallinity of the stretched samples [23]. In addition, biaxial stretching significantly decreases the permeability coefficient of the HDPE/GNP composite by two orders of magnitude due to the parallel alignment of GNPs in the stretching plane increasing the tortuous path of gaseous molecules, while the one-dimensional shape of CNTs and spherical shape of CB limit their effectiveness in improving the barrier properties of the polymer composites.

3.4 BIAXIALLY STRETCHED BINARY CARBON NANOFILLER-REINFORCED COMPOSITES

In this section, biaxial stretching of the composites containing 4 wt% binary carbon nanofillers (nanofiller weight ratio = 1:1) was conducted to investigate the effect of carbon nanofillers on material deformation behavior and the influence of biaxial deformation on the structure and properties of the deformed material. Furthermore, the possible synergistic effects between the two different types of carbon nanofillers after the orientation and deagglomeration of nanofillers were also explored in this study. In this work, the samples were sim-biaxially stretched at a strain rate of $4\,s^{-1}$ and a stretching temperature of 131°C with increasing SRs of 2, 2.5, and 3.

3.4.1 Structural Evolution

Figure 3.20 shows the structures of the unstretched and biaxially stretched HDPE/GNP/MWCNT, HDPE/CB/MWCNT, and HDPE/GNP/CB composites. It can be

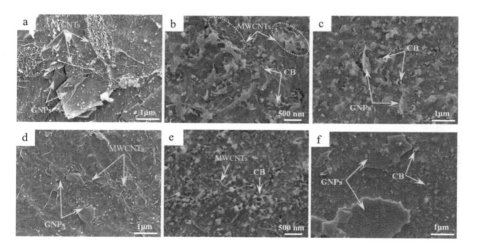

FIGURE 3.20 SEM images of unstretched and biaxially stretched binary carbon nanofiller-reinforced composites: (a,d) HDPE/GNP/MWCNT, (b,e) HDPE/CB/MWCNT, and (c,f) HDPE/GNP/CB. CB, carbon blank; GNP, graphene nanoplatelet; HDPE, high-density polyethylene; MWCNT, multi-walled carbon nanotube.

observed in Figure 3.20 that the morphological evolution of the carbon nanofillers in the matrix is not disrupted by the presence of another nanofiller during biaxial deformation. By way of example, the GNPs are aligned parallel to the surface of the stretched HDPE/GNP/MWCNT composite after biaxial stretching at an SR of 3, and the MWCNTs are disentangled and randomly oriented in the stretched sample, as shown in Figure 3.20a and b. By comparing the morphologies of the unstretched and stretched composites, it can be seen that the carbon nanofillers are dispersed more uniformly in the matrix for all the binary carbon nanofiller-reinforced composites after biaxial stretching.

The WAXRD spectra for the unstretched and stretched composites containing 4 wt% binary carbon nanofillers at different SRs are shown in Figure 3.21a–c. Figure 3.21d and e shows the changes in crystallinities and average crystallite sizes of the samples after biaxially stretching with increasing SRs, respectively. It can be observed in Figure 3.21a–c that the intensity in the (110) and (200) reflection planes for all the stretched samples decreases and a clear reflection peak in the (002) plane of graphene layers still exists for the stretched HDPE/GNP/MWCNT and HDPE/GNP/CB composites with different SRs. In Figure 3.21d, the crystallinity of all the biaxially stretched samples increases by about 8%–10%, indicating the crystallinity of the stretched samples is barely influenced by the addition of binary carbon nanofillers. However, the addition of binary carbon nanofillers results in the smaller average crystallite sizes for the stretched binary carbon nanofiller-reinforced composites, especially for the stretched HDPE/GNP/CB composites, as shown in Figure 3.21e. In Section 3.3.3, it was found that the GNPs and CB exhibit a strong restriction effect on the mobility of polymer chains, resulting in more imperfect crystallites.

3.4.2 Thermal Properties

The melting and crystallization behavior of the biaxially stretched binary carbon nanofiller-reinforced composites were investigated using DSC. The relevant thermal parameters obtained from the first heating and cooling stages are shown in Figure 3.22. It can be observed in Figure 3.22a that the crystallinity of all the stretched samples increases by about 10%–15%. In Figure 3.22b, a slight increase (about 1°C–2°C) in melting temperature can also be observed in the first heating stage for all the stretched samples due to the presence of crystallites with increased size. The decreased values in W_h for the stretched binary carbon nanofiller-reinforced composites indicate a narrower distribution of crystallite sizes compared to the stretched neat HDPE samples (Figure 3.22c). It can be seen in Figure 3.22d that the crystallization temperatures of all the neat HDPE and binary composite samples are also not significantly influenced by biaxial stretching. Overall, there is no significant difference in the thermal properties between the stretched unary and binary carbon nanofiller-reinforced composites.

3.4.3 Tensile Properties

Figure 3.23 shows the tensile properties of the biaxially stretched binary carbon nanofiller-reinforced composites [24]. The effect of the addition of 4 wt% binary carbon

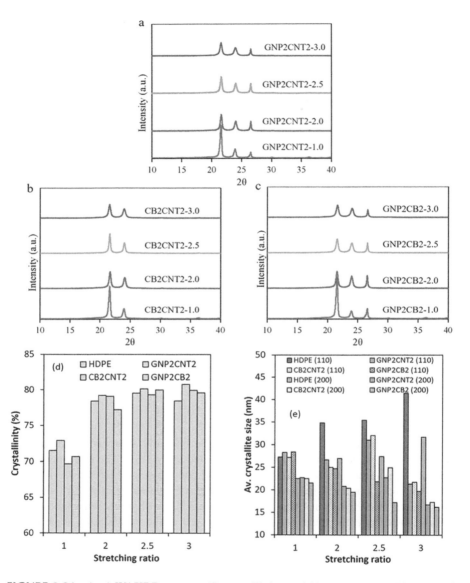

FIGURE 3.21 (a–c) WAXRD spectra, (d) crystallinity, and (e) average crystallite sizes of binary carbon nanofiller-reinforced composite sheets with increasing SRs.

nanofillers on the tensile properties of biaxially stretched samples with increasing SRs is shown in Table 3.5. It can be observed in Figure 3.23a that the modulus of the stretched composites containing 4 wt% binary carbon nanofillers increases steadily with increasing SRs due to the deagglomeration and orientation of nanofillers. The modulus of HDPE/GNP/MWCNT composites is clearly higher than the modulus of HDPE/CB/MWCNT and HDPE/GNP/CB composites at all the SRs. The predicted modulus from the Halpin–Tsai model ($E_b = E_m + \Delta E_1 + \Delta E_2$) for the binary carbon

Biaxially Stretched Conductive Polymer Composites

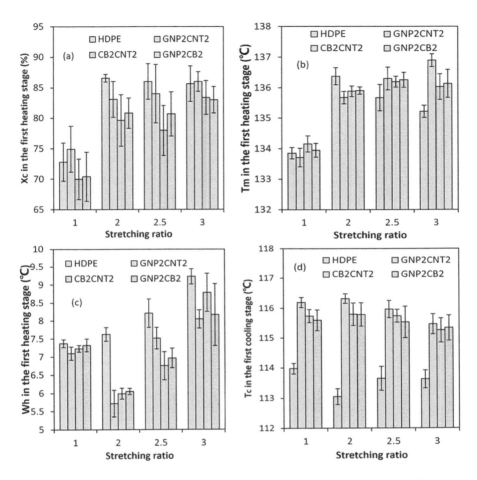

FIGURE 3.22 Thermal parameters of the biaxially stretched binary carbon nanofiller-reinforced composites with increasing SRs: (a) crystallinity, (b), melting temperature, (c) width at half height, (d) crystallization temperature. SRs, stretching ratios.

nanofiller-reinforced composites at different SRs is also shown in Figure 3.23a. It should be noted that the contribution to modulus from the GNPs in the stretched HDPE/GNP/MWCNT and HDPE/GNP/CB composites was predicted by the parallel orientation Halpin–Tsai model $\left(\dfrac{E}{E_m} = \dfrac{1+\xi\eta\varphi_f}{1-\eta\varphi_f}\right)$, while the contribution to modulus from the MWCNTs and CB was predicted by the random orientation Halpin–Tsai model $\left(\dfrac{E_r}{E_m} = \dfrac{3}{8}\left[\dfrac{1+\xi\eta_L\varphi_f}{1-\eta_L\varphi_f}\right] + \dfrac{5}{8}\left[\dfrac{1+2\eta_T\varphi_f}{1-\eta_T\varphi_f}\right]\right)$, according to the morphological characteristics of the stretched binary carbon nanofiller-reinforced composites in Figure 3.20. It can be seen from Figure 3.23a that the predicted modulus for all the stretched HDPE/GNP/MWCNT and HDPE/GNP/CB composites is higher than the experimental values. This may indicate that the parallel orientation Halpin–Tsai model is

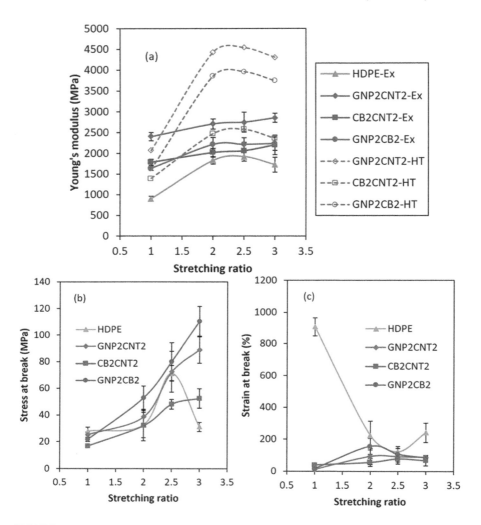

FIGURE 3.23 Effect of biaxial stretching on the (a) Young's modulus, (b) stress at break, and (c) strain at break of binary carbon nanofiller-reinforced composites with increasing SRs. SRs, stretching ratios.

not suitable for the stretched composites containing GNPs in this case. The errors between the predicted modulus and experimental modulus for the stretched HDPE/CB/MWCNT composite are much smaller.

It can be seen in Figure 3.23b that the stress at break of all the stretched binary carbon nanofiller-reinforced composites is significantly increased, especially for the stretched HDPE/GNP/CB composites. The strain at break of the binary composite samples is also increased due to the deagglomeration of nanofillers during biaxial deformation, as shown in Figure 3.23c.

One can see in Table 3.5 that the HDPE/GNP/MWCNT composite still exhibits the most efficient reinforcement in modulus after biaxial stretching due to the high

TABLE 3.5
Effect of the Addition of 4 wt% Binary Carbon Nanofillers on the Tensile Properties of Samples Sim-Biaxially Stretched at a Strain Rate of 4 s⁻¹ and a Stretching Temperature of 131°C

Sample	SR	ΔE (%)	$\Delta\sigma_b$ (%)	$\Delta\varepsilon_b$ (%)
GNP2CNT2	2	+48.1	+18.5	−58.0
CB2CNT2		+10.5	−1.7	−74.5
GNP2CB2		+21.1	+62.5	−30.2
GNP2CNT2	2.5	+43.2	+1.5	−24.7
CB2CNT2		+7.4	−32.6	−35.7
GNP2CB2		+15.2	+12.2	−12.0
GNP2CNT2	3	+65.3	+186.1	−63.3
CB2CNT2		+27.6	+68.8	−71.7
GNP2CB2		+29.4	+255.6	−64.2

CB, carbon blank; CNT, carbon nanotube; GNP, graphene nanoplatelet; SR, stretching ratio.

aspect ratios of GNPs and MWCNTs, while the HDPE/GNP/CB composite exhibits the most efficient reinforcement in stress at break after biaxial deformation due to the high aspect ratio of GNPs and the intensive deagglomeration of CB. At an SR of 3, the E of the HDPE/GNP/MWCNT composite increases by about 65% and the σ_b of the HDPE/GNP/CB composite increases by about 256%, compared with the unfilled HDPE. By comparing the effect of 4 wt% unary and binary carbon nanofillers on the tensile properties of biaxially stretched samples (Tables 3.4 and 3.5), it is found that there is no evident synergistic effect between the binary carbon nanofillers in the tensile properties of the stretched samples.

3.4.4 ELECTRICAL PROPERTIES

The variations in the volume resistivity of the biaxially stretched composites with 4 wt% binary carbon nanofillers with increasing SRs are shown in Figure 3.24. It can be observed in Figure 3.24 that the stretched HDPE/GNP/CB composites at all the SRs level off at a high resistivity.

The resistivity of the HDPE/GNP/MWCNT composites increased by nine to ten orders of magnitude after biaxial stretching, although the parallel GNPs appear to be bridged by the randomly oriented MWCNTs in Figure 3.20d. This can be attributed to the distance between the GNPs and MWCNTs being still higher than the critical maximum distance for electron hopping after biaxial deformation. Also, the resistivity of the stretched HDPE/CB/MWCNT composites increases significantly due to the destruction of conductive pathways, as shown in Figure 3.24. Wen et al. [25] reported a grape-cluster-like conductive network in a PP/CB/MWCNT composite prepared by multistage stretching extrusion, where the oriented MWCNTs (branch-like) provided charge transport over large distances, while the grape-like CB clusters

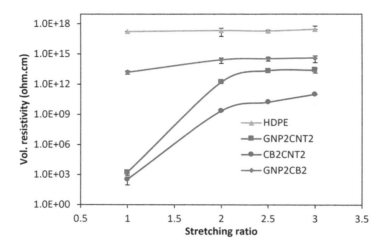

FIGURE 3.24 Variations in the volume resistivity of biaxially stretched binary carbon nanofiller-reinforced composites as a function of SRs. SRs, stretching ratios.

bridged the MWCNTs through charge transport over small distances. However, this grape-cluster-like conductive network was not generated in this study, which may be attributed to the grape-like CB clusters (or high structure CB) being deagglomerated during biaxial deformation (Figure 3.20b).

Comparing the resistivity of the stretched binary carbon nanofiller-reinforced composites (Figure 3.24) with that of the stretched composite containing 4 wt% MWCNTs (Figure 3.10) under the same stretching conditions, one can see that the MWCNT-filled composites exhibit less increase in resistivity with increasing SRs. Therefore, it is found from this work that it is very difficult to avoid the destructive effect of biaxial stretching on the conductive network even for mixed particle systems.

3.4.5 BARRIER PROPERTIES

The oxygen permeability coefficient (P_g) of the unstretched and biaxially stretched samples at an SR of 3 was measured to investigate the effect of binary nanofillers and biaxial stretching on barrier properties, as shown Figure 3.25 [24]. The P_g of unstretched HDPE/CB/CNT composite shows a slight decrease of around 10% compared with the neat HDPE. However, the P_g of both unstretched HDPE/GNP/CNT and HDPE/GNP/CB composites significantly decreased by about 35% due to the presence of GNPs that have a very high aspect ratio and typical two-dimensional structure resulting in an increased tortuosity in the diffusion path of gaseous molecules. The P_g of the stretched HDPE and HDPE/CB/CNT composite decreased by ~10% probably due to increased crystallinity of the stretched samples [23]. Importantly, the P_g of stretched HDPE/GNP/CNT and HDPE/GNP/CB composites decreased by 88% and 97%, respectively, compared with the unstretched ones, demonstrating that the parallel alignment of GNPs to the stretching surface improves the barrier property of the composites. The spherical CB served as spacers may facilitate the

Biaxially Stretched Conductive Polymer Composites

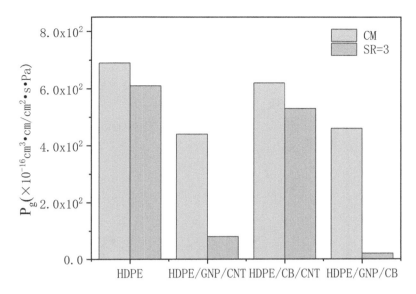

FIGURE 3.25 Effect of biaxial stretching on the permeability coefficient of the composites.

realignment and deagglomeration of GNPs in the HDPE/GNP/CB composite upon biaxial stretching, resulting in the lowest P_g (see the schematic "tortuous path" for the HDPE/GNP composite in Section 3.4.6).

3.5 BIAXIAL STRETCHING OF PP/MWCNT AND TPU/MWCNT/RGO COMPOSITES

This chapter also summarizes several studies on biaxially stretched PP/MWCNT and TPU (thermoplastic polyurethane)/MWCNT/rGO (reduced graphene oxide) composites. For instance, Shen et al. [1,9] investigated the development of carbon nanotube network in PP/CNT nanocomposites, reveling a special process of destruction and rebuilding of conductive network during simultaneous biaxial stretching. They found that during the sim-biaxial stretching (SIB stretching), the conductivity of composite decreased at the initial stretching, while the conductivity began to increase with the SR increased beyond 2.5×2.5 (Figure 3.26). A similar trend was found in the recently published biaxially stretched TPU/rGO and TPU/MWCNT/rGO composites [26], as shown in Figure 3.27. However, the rebuilding process of conductive networks was not observed in the studies on biaxially stretched CNT composites with some other polymer matrices, such as PET/MWCNT and HDPE/MWCNT [4,8]. There are many possible reasons for these differences in the structural evolution and final electrical and mechanical properties of samples, such as polymer matrix, strain rate, stretching mode, and stretching temperature. More work is required to achieve a better understanding of structure evolution and properties of polymer/carbon nanofiller composites under biaxial tension.

Recently, rGO decorated with immobilized CNTs was introduced into a TPU matrix to obtain a conductive nanocomposite, then the obtained TPU/CNT/rGO

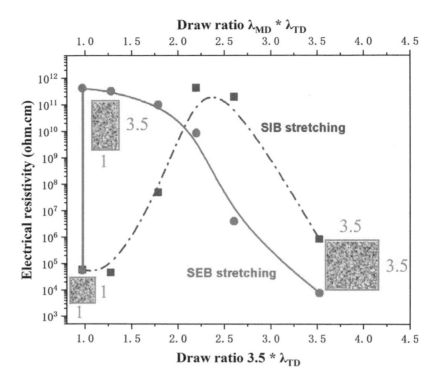

FIGURE 3.26 Electrical resistivities of iPP/CNT composites containing 2.2 vol.% CNTs as a function of draw ratios during sim-biaxial (SIB) and seq-biaxial (SEB) stretching.

sheets sequentially biaxially stretched to produce thin flexible films for sensing applications [26].

The improvement of strain sensing performance of flexible strain sensors indicates that the biaxial stretching process is beneficial to promote the dispersion of nanofillers in the TPU and reconstruct the conductive network resulting in a sensor with enhanced sensitivity compared to the unstretched sensors. For instance, a high sensitivity (Gauge factor/GF = 150 at 30% strain) obtained in the TPU/rGO$_{4\times4}$ composite with an SR of 4 in each direction, but the sensor has a limited strain range detection (0%–150%) and low linearity ($R^2 = 0.76$ at a strain of 30%). The TPU/CNT/rGO$_{4\times4}$ sample is less sensitive (GF = 42.3 at 30% strain) than TPU/rGO$_{4\times4}$, but has a higher linearity ($R^2 = 0.91$ at a strain of 30%) and a larger detectable range (0%–400%). This is due to the fact that while the rGO sheets are prone to slippage under an external force, inducing irreversible destruction to the conductive network structure, the addition of CNTs improves the resistance of the conductive network to deformation through a synergy with rGO [27] (Figure 3.28).

The strain sensor designed in the work was found to be suitable for the detection of physiological activities and human body health, such as touch sensing, finger movement, heart rate recognition, exhalation, swallowing, and language recognition (Figure 3.29) [26].

Biaxially Stretched Conductive Polymer Composites

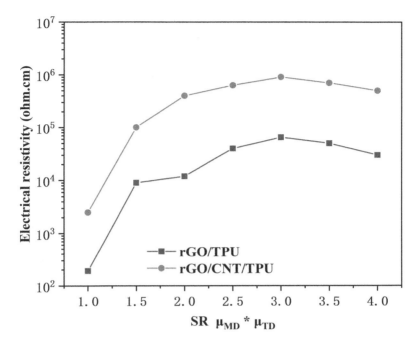

FIGURE 3.27 Electrical resistivity of TPU/rGO and TPU/CNT/rGO composite changes as a function of SR. rGO, reduced graphene oxide; TPU, thermoplastic polyurethane.

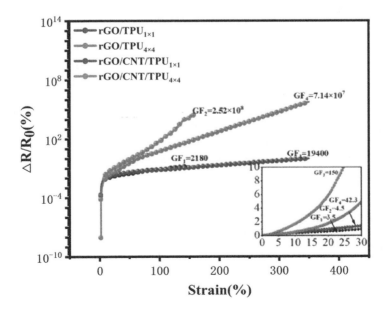

FIGURE 3.28 Relative resistance change ($\Delta R/R_0$) of unstretched and biaxially stretched composites as a function of strain.

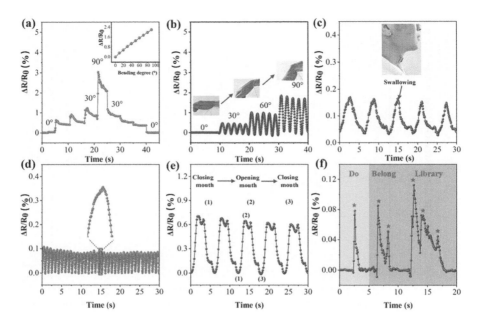

FIGURE 3.29 Electromechanical responses of the sensor to (a) index finger and (b) wrist bending/unbending; (c) swallowing; (d) heart rate recognition; (e) mouth opening; and (f) speaking "Do", "Belong", and "Library".

3.6 CONCLUSIONS

Compared with traditional methods, composite sheets prepared via biaxial stretching have advantages in strength, rigidity, stability, optical properties, thickness uniformity, and so on. In this chapter, the effect of biaxial stretching on the properties of polymer composites with conductive carbon nanofillers has been summarized. A suitable SR and careful choice of nanofiller can provide polymer nanocomposite materials with a unique structure and morphology as well as a high orientation degree to enhance mechanical, electrical, and barrier properties. As multifunctional high-performance materials, biaxially stretched polymer/carbon nanofiller nanocomposites have a significant potential for application in the fields of packaging and strain sensors, etc.

REFERENCES

1. Shen JB, Champagne MF, Gendron R, Guo S. The development of conductive carbon nanotube network in polypropylene-based composites during simultaneous biaxial stretching. *Eur Polym J.* 2012;48(5):930–9.
2. Shen YC, Harkin-Jones E, Hornsby P, McNally T, Abu-Zurayk R. The effect of temperature and strain rate on the deformation behaviour, structure development and properties of biaxially stretched PET-clay nanocomposites. *Compos Sci Technol.* 2011;71(5):758–64.
3. Soon KH, Harkin-Jones E, Rajeev RS, Menary G, McNally T, Martin PJ, et al. Characterisation of melt-processed poly(ethylene terephthalate)/synthetic mica nanocomposite sheet and its biaxial deformation behaviour. *Polym Int.* 2009;58(10):1134–41.

4. Xiang D, Harkin-Jones E, Linton D. Processability, structural evolution and properties of melt processed biaxially stretched HDPE/MWCNT nanocomposites. *RSC Adv.* 2014;4(83):44130–40.
5. Xiong J, Wang X, Zhang X, Xie Y, Lu J, Zhang Z. How the biaxially stretching mode influence dielectric and energy storage properties of polypropylene films. *J Appl Polym Sci.* 2020;138(11):50029.
6. Chen Q, Wang Z, Zhang S, Cao Y, Chen J. Structure evolution and deformation behavior of polyethylene film during biaxial stretching. *ACS Omega.* 2020;5(1):655–66.
7. Onyishi HO, Oluah CK. Effect of stretch ratio on the induced crystallinity and mechanical properties of biaxially stretched PET. *Phase Transit.* 2020;93(9):924–34.
8. Mayoral B, Hornsby PR, McNally T, Schiller TL, Jack K, Martin DJ. Quasi-solid state uniaxial and biaxial deformation of PET/MWCNT composites: structural evolution, electrical and mechanical properties. *RSC Adv.* 2013;3(15):5162–83.
9. Shen JB, Champagne MF, Yang Z, Yu Q, Gendron R, Guo S. The development of a conductive carbon nanotube (CNT) network in CNT/polypropylene composite films during biaxial stretching. *Compos Part A-Appl S.* 2012;43(9):1448–53.
10. Xiang D, Harkin-Jones E, Linton D. Characterization and structure-property relationship of melt-mixed high density polyethylene/multi-walled carbon nanotube composites under extensional deformation. *RSC Adv.* 2015;5(59):47555–68.
11. Xu D, Wang Z. Role of multi-wall carbon nanotube network in composites to crystallization of isotactic polypropylene matrix. *Polymer.* 2008;49(1):330–8.
12. Luo W, Zhou N, Zhang Z, Wu H. Effects of vibration force field on structure and properties of HDPE/CaCO$_3$ nanocomposites. *Polym Test.* 2006;25(1):124–9.
13. Zhao X, Ye L. Structure and properties of highly oriented polyoxymethylene/multi-walled carbon nanotube composites produced by hot stretching. *Compos Sci Technol.* 2011;71(10):1367–72.
14. Yang J, Wang K, Deng H, Chen F, Fu Q. Hierarchical structure of injection-molded bars of HDPE/MWCNTs composites with novel nanohybrid shish-kebab. *Polymer.* 2010;51(3):774–82.
15. Li CY, Thostenson ET, Chou T-W. Dominant role of tunneling resistance in the electrical conductivity of carbon nanotube-based composites. *Appl Phys Lett.* 2007;91(22):223114(1–3).
16. Socher R, Krause B, Müller MT, Boldt R, Pötschke P. The influence of matrix viscosity on MWCNT dispersion and electrical properties in different thermoplastic nanocomposites. *Polymer.* 2012;53(2):495–504.
17. Huang YY, Terentjev EM. Dispersion of carbon nanotubes: Mixing, sonication, stabilization, and composite properties. *Polymers-Basel.* 2012;4(1):275–95.
18. Alig I, Skipa T, Lellinger D, Pötschke P. Destruction and formation of a carbon nanotube network in polymer melts: Rheology and conductivity spectroscopy. *Polymer.* 2008;49(16):3524–32.
19. Xiang D, Wang L, Zhang Q, Chen B, Li Y, Harkin-Jones E. Comparative study on the deformation behavior, structural evolution, and properties of biaxially stretched high-density polyethylene/carbon nanofiller (carbon nanotubes, graphene nanoplatelets, and carbon black) composites. *Polym Composite.* 2018;39(S2):E909–E23.
20. Abu-Zurayk R, Harkin-Jones E, McNally T, Menary G, Martin P, Armstrong C. Biaxial deformation behavior and mechanical properties of a polypropylene/clay nanocomposite. *Compos Sci Technol.* 2009;69(10):1644–52.
21. Du J, Zhao L, Zeng Y, Zhang L, Li F, Liu P, et al. Comparison of electrical properties between multi-walled carbon nanotube and graphene nanosheet/high density polyethylene composites with a segregated network structure. *Carbon.* 2011;49(4):1094–100.
22. Deng H, Zhang R, Bilotti E, Loos J, Peijs T. Conductive polymer tape containing highly oriented carbon nanofillers. *J Appl Polym Sci.* 2009;113(2):742–51.

23. Chatterjee T, Patel R, Garnett J, Paradkar R, Ge S, Liu L, et al. Machine direction orientation of high density polyethylene (HDPE): Barrier and optical properties. *Polymer.* 2014;55(16):4102–15.
24. Xiang D, Wang L, Tang YH, Harkin-Jones E, Zhao CX, Li YT. Processing-property relationships of biaxially stretched binary carbon nanofiller reinforced high density polyethylene nanocomposites. *Mater Lett.* 2017;209:551–4.
25. Wen M, Sun X, Su L, Shen J, Li J, Guo S. The electrical conductivity of carbon nanotube/carbon black/polypropylene composites prepared through multistage stretching extrusion. *Polymer.* 2012;53(7):1602–10.
26. Zhang X, Xiang D, Wu Y, Harkin-Jones E, Shen J, Ye Y, et al. High-performance flexible strain sensors based on biaxially stretched conductive polymer composites with carbon nanotubes immobilized on reduced graphene oxide. *Compos Part A-Appl S.* 2021;151:106665.
27. Cai WT, Huang Y, Wang DY, Liu CX, Zhang YG. Piezoresistive behavior of graphene nanoplatelets/carbon black/silicone rubber nanocomposite. *J Appl Polym Sci.* 2014;131(3):39778.

4 Blown Film Extrusion of Conductive Polymer Composites

4.1 INTRODUCTION

Blown film extrusion is an important polymer manufacturing process [1–3] used to produce commodity and specialized polymer films, typically used in packagings such as shrink, stretch, barrier films (used to protect deli meat), frozen food packaging, and shopping bags. The schematic diagram for the blown film extrusion is shown in Figure 4.1 [4]. The blown film extrusion roughly includes the following processes: the first step involves melting the polymer in an extruder [5]. Polymer resin, often in the form of beads, is loaded into a hopper and fed into a heated barrel with a screw that transports the polymer down the barrel. Beads are gradually heated to melt the polymer. The heat profile is an important part of extrusion since the polymers are temperature sensitive. When the molten material reaches the end of the barrel [6], it is extruded through a die. This point differentiates blown film extrusion from other extrusion processes [7]. The molten polymer enters the die head, and the air is injected via a hole in the die center to radially inflate the polymer into a thin tube. This step is adjusted to achieve the desired film thickness and width.

The blow-up ratio (BUR) and the draft ratio are two important parameters in blown film extrusion. The BUR is the ratio of the diameter of the membrane tube after blowing to the diameter of the annular die. Whereas, the draft ratio is the ratio of the traction speed of the traction roller to the melt extrusion speed at the die mouth. During blowing and stretching processes, the molecules show a directional effect, and the draft ratio in the longitudinal or machine direction (MD) and the BUR in the transverse direction (TD) are always equal. The film with the best physical and mechanical properties is obtained by correctly selecting the BUR and drafting ratio. In practice, films with different thicknesses are often obtained with the same die and different traction speeds; therefore, the longitudinal and transverse strength of films is often different.

The hot tube film is then cooled, typically with high-speed air, and pulled upward by nip rollers. In the medium-to-large-size film lines, this vertical tube extends several stories into the air. When the film crystallizes after cooling, it is called the frost line. As the tube reaches the nip rollers, at the top of the line, the film is cooled enough to flatten and then referred to as lay-flat or collapsed. The film is then transported downstream by rollers for further processing, such as slit, printed, vented, converted into bags, and eventually wound into rolls.

Blown film extrusion is a polymer processing technique to efficiently produce thin polymer films. Although several billion pounds of polymers are processed into films by blown film extrusion every year, and this technique has been investigated for pure

DOI: 10.1201/9781003218661-4

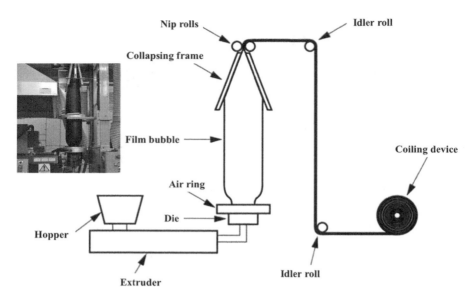

FIGURE 4.1 The schematic diagram of the blown film extrusion forming process.

polymers in some previous articles [8], however, there are very few studies available on the blown film extrusion of polymer/carbon nanofiller composites. Therefore, the research on blown film extrusion of polymer/carbon nanofiller composites is extensively reviewed in this chapter [9].

4.2 BLOWN FILM EXTRUSION OF THERMOPLASTIC POLYURETHANE/CNTs

Thermoplastic polyurethane (TPU) is a linear block copolymer characterized by hard segments (HSs) and soft segments (SSs). HSs are made from diisocyanate, while the SSs consist of long flexible polyether or polyester chains that interconnect two HSs. In particular, the HSs act as multifunctional tie points working as both physical crosslinks and reinforcing fillers, while the SSs primarily influence the elastic properties of TPU. This special chemical structure imparts very interesting characteristics, such as a wide range of operating temperature and harness options, excellent wear and tear resistance, good non-polar solvent resistance, high compression, and high tensile strength. To improve and develop the application field of TPU, some specific properties such as mechanical or dielectric properties can be adjusted by adding nanofillers. P. Russo et al. [10] used CNT as reinforcing fillers in TPU composites to investigate the effects of different processing conditions on the properties of blown extruded composite films.

4.2.1 MORPHOLOGY

The dispersion of nanofillers is usually affected by aggregation, which determines the reduction of their effectiveness. Figure 4.2 compares cryo-fractured surface

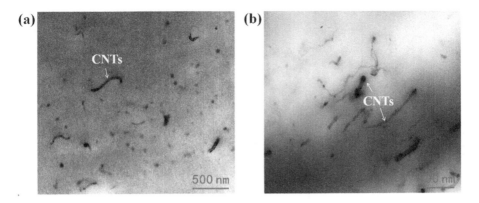

FIGURE 4.2 Comparison of TEM micrographs of TPU-based film samples containing 1 wt% CNTs and obtained by (a) chill-roll extrusion and (b) film blowing technologies.

TABLE 4.1
Interdomain Spacing and Phase Separation Degree Values of TPU and Nanocomposites

Preparation Method	Sample (CNT Content)	Interdomain Spacing (nm)	Phase Separation Degree
Blown film extrusion	EL 1185A (0 wt%)	7.50	0.740
	0.2 wt%	8.32	0.758
	0.5 wt%	8.66	0.760
	1 wt%	8.88	0.770

images (at the same magnification) of polymers prepared using (a) chill-roll extrusion and (b) film blowing technology. In terms of distribution of fillers, the chill-roll extrusion technology ensures that the dispersion of the contained nanotubes is better than the blowing conditions.

4.2.2 STRUCTURE

The degree of phase separation values is reported in Table 4.1. For blown extruded composite films, the domain spacing increases with multi-walled carbon nanotube (MWCNT) content. This is in good agreement with the findings of Xia et al [11]. Figure 4.3 shows the interdomain spacing values as a function of the degree of phase separation for TPU and the related nanocomposite films. The domain spacing of blown extruded composite films increases linearly with the degree of phase separation (DPS). In particular, the higher the DPS, the higher the distance between the hard and soft domains. An increase in CNT content leads to an increase in hard domain size, regardless of the processing technology [10].

Different morphologies observed for nanocomposite film samples are ascribed to the effect of different cooling rates on the mechanism of nucleation, growth, and

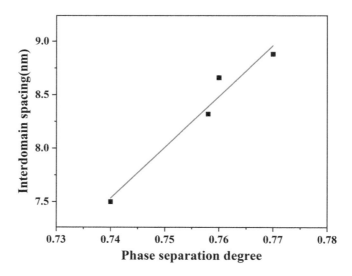

FIGURE 4.3 Interdomain spacing as a function of the degree of phase separation for neat TPU and nanocomposites prepared by film blowing.

TABLE 4.2
Tensile Parameters of All Investigated Film Samples

Sample (CNT content)	E (MPa)	ε_{br} (%)	σ_{max} (MPa)
EL 1185A (0 wt%)	10.1 ± 1.0	730 ± 23	24.3 ± 3.8
0.2 wt%	12.6 ± 2.8	574 ± 57	9.7 ± 2.1
0.5 wt%	13.2 ± 4.1	627 ± 56	10.4 ± 1.7
1 wt%	11.1 ± 1.7	608 ± 21	11.9 ± 1.1

self-assembly of hard domains. At relatively low cooling rates, the HSs maintain sufficient mobility for a longer time in film blowing technology. Thus the HSs, as well as the nucleated hard domains, may further self-assemble to form larger hard domains. About 0.5 wt% CNT nanocomposite is annealed at 85°C for 15 minutes, and the results reveal that low cooling rates allow the growth of small hard domains into larger domains through a self-assembling mechanism.

4.2.3 Tensile Properties

For the tensile behavior, mainly longitudinal studies, the properties of pure TPU film and CNT film are shown in Table 4.2. With the increase in CNT content, the tensile stiffness is improved, and the tensile strength is halved; however, the effect of elongation at break is not obvious, increasing slightly from 0.2 to 1 wt%. This is explained by the balance between the stiffening due to the filler and the orientation and/or organization of the dispersed phase in the amorphous regions and the TPU softening for the mutual organization of the hard matrix segments and domains.

TABLE 4.3
Dynamic-Mechanical Parameters for All Investigated Film Samples

Samples (MWCNT Content)	$E'_{-80°C}$ (MPa)	% Variation with Plain Samples	$E'_{-40°C}$ (MPa)	% Variation with Plain Samples
EL 1185A (0 wt%)	1,041		10.6	
0.2 wt%	1,170	12	19.8	92
0.5 wt%	1,172	13	22.1	108
1.0 wt%	1,722	65	24.3	129

4.2.4 Dynamic-Mechanical Properties

The blown film samples filled with CNTs show a hardening effect in a wide temperature range, especially in the rubber region. The 0.5 wt% CNTs content is not sufficient to improve the storage modulus of the film in the glass area; however, it is effective only in the rubber area. Doubling the filler content significantly improves the storage modulus. Table 4.3 shows the storage modulus values of blown film samples in the glass area and rubber area.

The analysis of the viscoelastic behavior confirms the reorganization of different phases dictated by the rate of cooling of the film and the CNT content. For blown films, the slow cooling allows the aggregation of hard domains that, in the final structure, increase dimensions with increasing nanotubes content. This is in good agreement with the morphological data, indicating that the increase in CNT concentration increases the storage modulus over the entire range of temperatures for 1% filler content. For lower filler contents, the same parameter shows an improvement only in the rubbery region where this effect is interpreted as a greater consistency of the amorphous phase due to the presence of rigid nanoparticles.

4.3 BLOWN FILM EXTRUSION OF TPU/GRAPHENE OXIDE NANOCOMPOSITES

TPU is widely used in various applications, such as fibers, coatings, adhesives, and biomedical projects, due to its melt processability and multifunctional characteristics closely related to the inherent two-phase segmented structure. However, its low stiffness, tensile strength, and weak barrier still limit its large-scale practical applications. At present, most of them change the separation hard/soft region of the TPU matrix by adding nanofillers, and remarkable results have been achieved. Russo et al. [12] focused on the blown film based on TPU composites. By adding graphene oxide (GO), the structure of the TPU matrix is modified, thereby, affecting the water vapor barrier performance.

4.3.1 Thermal Properties

Thermal analysis of TPU/GO blown film and pure TPU blown film samples is routinely prepared by typical film blowing technology. Table 4.4 summarizes the

TABLE 4.4
Calorimetric Parameters of Investigated Blown Film

Sample	$T_{m,s}$ (°C)	$\Delta H_{m,s}$ (J/g)	$T_{m,h}$ (°C)	$\Delta H_{m,h}$ (J/g)
TPU+0%GO	127.32	4.54	167.09	6.07
TPU+0.2%GO	107.51	1.36	168.45	4.84
TPU+0.5%GO	99.34	1.41	165.23	8.50
TPU+1%GO	103.44	1.59	165.80	7.09

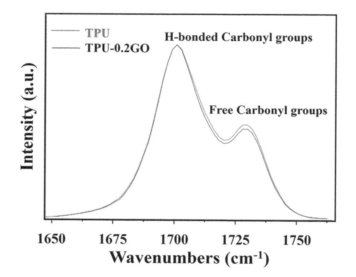

FIGURE 4.4 FT-IR spectra of pure TPU blown film and TPU/GO blown film.

temperature and enthalpy of TPU/GO composites. Since TPU contains soft phase and hard phase structure, it generates a complex thermal spectrum. The hard and soft phases in TPU are semi-crystalline; however, the crystallinity is not very high. The melting temperature of the soft phase is lower than that of the hard phase. GO content has little effect on the melting temperature of the hard phase. Most significant changes are detected by the enthalpies of melting with relevant reductions in thermal parameters for the soft phase of nanocomposite systems.

4.3.2 STRUCTURE

To quantify the separation of hard domains, TPU/GO blown film containing 0.2 wt% GO and pure TPU blown film is analyzed by Fourier transform infrared spectroscopy (FT-IR) spectroscopy, and carbonyl peaks in the region of 1,650–1,775 cm^{-1} are studied (Figure 4.4). The absorption peaks at 1,730, 1,712, and 1,700 cm^{-1} are assigned to free carbonyl (i.e., non-hydrogen bond), involving hydrogen bond carbonyl of disordered hard domains and strong hydrogen bond carbonyl of ordered hard domains, respectively.

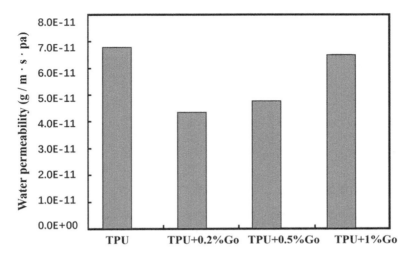

FIGURE 4.5 Water permeability for pristine TPU blown film and TPU/GO blown film.

The degree of phase separation (DPS) is estimated by the ratio of the absorption area of the peak assigned to the bonded carbonyl group to the peak area assigned to the free and bonded carbonyl group. According to this definition, the DPS (a measure of the DPS) increases slightly due to the presence of GO nanoparticles, attributed to the nucleation effect of the graphite layer, which promotes the orientation of TPU macromolecules, thus establishing the hydrogen network and the possible interaction between the carbonyl group of TPU and the hydroxyl groups on the GO layer.

4.3.3 Barrier Properties

Water vapor permeability (WVP) measurements are determined by the infrared sensor method. To study the effect of GO in the blown film extrusion process on the barrier property of the TPU matrix under different loading, a WVP test is carried out. The WVP results, shown in Figure 4.5, suggest that the presence of a small amount of (0.2% wt) GO significantly reduces the permeability by 36%. The existence of layered GO fillers increases the tortuous paths of diffusing agent molecules and increases the hard domains. In contrast, the higher the filler content increases the water permeability of TPU + 0.2 wt% GO sample, due to the uneven dispersion of GO filler in the TPU matrix.

4.4 BLOWN FILM EXTRUSION OF HIGH-DENSITY POLYETHYLENE/CNT COMPOSITES

Xiang et al. [13] introduced film blowing extrusion to pure HDPE and HDPE/MWCNT composite films, and studied the thermal, mechanical, and electrical properties of films. In addition, the effect of annealing on the electrical properties of HDPE/MWCNT (4 wt%) composite film was also studied.

4.4.1 Morphology

SEM images of HDPE/MWCNT composite blown films containing 4 wt% MWCNTs at BURs of 2 and 3 are shown in Figure 4.6a and b, respectively. MWCNTs oriented along the flow direction and randomly oriented are observed at a BUR of 2 [14]. On the other hand, there is less clear MWCNT orientation along the flow direction and fewer MWCNT agglomerates due to the increased deformation in the TD at a BUR of 3. The disentangled MWCNTs in the film with a BUR of 3 are deagglomerated after annealing at 140°C for 2 hours, as shown in Figure 4.6c. The reagglomeration of nanotubes in polymers after annealing was also observed by Alig [15] and Jiang [16].

4.4.2 Thermal Properties

The melting and crystallization patterns of the blown films are investigated using DSC, and the results are compared with those of compression molded samples. The first heating and cooling curves and relevant thermal parameters are shown in Figure 4.7 and Table 4.5, respectively.

Figure 4.7a–c reveals that the melting peaks of neat HDPE and HDPE/MWCNT composites shift to a slightly lower temperature, relative to the corresponding compression molded samples, due to the decrease in crystallite size [17,18]. Figure 4.7d suggests

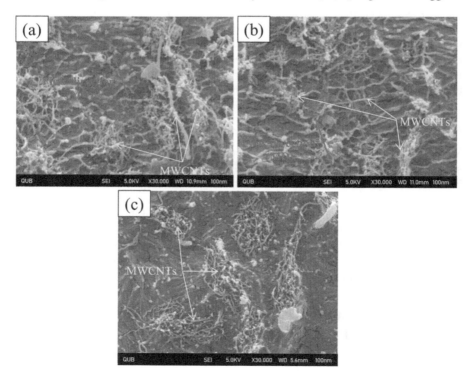

FIGURE 4.6 SEM images of the blown films of HDPE/MWCNT composite containing 4 wt% MWCNTs: (a) BUR2, (b) BUR3, and (c) BUR3, annealed at 140°C for 2 hours.

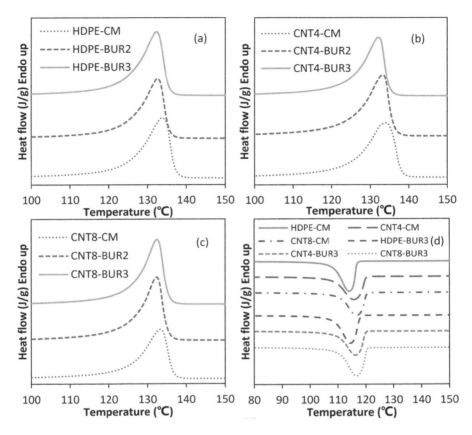

FIGURE 4.7 DSC curves of blown films of neat HDPE and HDPE/MWCNT composites with different BURs in the first heating (a–c) and cooling (d) stages.

TABLE 4.5
Thermal Parameters of the Compression Molded Samples and Blown Films of Pure HDPE and HDPE/MWCNT Composites with Different BURs

Sample	X_c^{1st} (%)	T_m^{1st} (°C)	W_h^{1st} (°C)	T_c (°C)	X_c^{2nd} (%)	T_m^{2nd} (°C)
HDPE-CM	72.8 ± 3.1	133.9 ± 0.2	7.4 ± 0.1	114.0 ± 0.2	70.4 ± 1.3	134.1 ± 0.1
CNT4-CM	73.9 ± 5.1	134.0 ± 0.1	7.7 ± 0.3	116.0 ± 0.2	70.7 ± 2.4	134.3 ± 0.2
CNT8-CM	70.9 ± 3.5	133.6 ± 0.3	7.1 ± 0.3	116.2 ± 0.4	72.6 ± 1.9	133.9 ± 0.4
HDPE-BUR2	72.3 ± 2.9	132.8 ± 0.2	6.6 ± 0.2	114.2 ± 0.2	70.1 ± 2.5	134.5 ± 0.2
HDPE-BUR3	73.7 ± 1.4	132.4 ± 0.1	6.6 ± 0.1	114.2 ± 0.4	71.5 ± 1.3	134.5 ± 0.1
CNT4-BUR2	76.6 ± 3.6	133.0 ± 0.7	6.8 ± 0.1	115.5 ± 0.4	75.2 ± 3.0	135.3 ± 0.1
CNT4-BUR3	77.2 ± 3.0	132.3 ± 0.2	6.6 ± 0.4	116.0 ± 0.3	74.7 ± 1.1	134.7 ± 0.5
CNT8-BUR2	76.6 ± 1.5	132.4 ± 0.2	5.7 ± 0.3	116.3 ± 0.1	75.1 ± 2.7	134.0 ± 0.2
CNT8-BUR3	76.6 ± 1.4	132.6 ± 0.2	5.9 ± 0.2	116.5 ± 0.2	75.8 ± 2.8	134.9 ± 0.2

that the crystallization behavior of the filled and unfilled samples is not influenced by the blown film extrusion process. Table 4.5 shows that the melting temperature in the first heating stage $\left(T_m^{1st}\right)$ at BURs of 2 and 3 decreases by 1°C–2°C. The X_c^{1st} of the blown films of pure HDPE does not change compared with the compression molded sample though a strain-induced crystallization is expected in the blown film, while that of HDPE/MWCNT composites exhibits a slight increase by 5% after blown film extrusion, attributed to a more efficient heterogeneous nucleation of MWCNTs in the blown film due to disentanglement of CNTs under elongational deformation. The nucleation effect of MWCNTs is observed in the cooling stage of the DSC tests, where the crystallization temperature (T_c) of all composites with 4 and 8 wt% MWCNTs increases by ~2°C, compared to pure HDPE. The crystallization behavior of the filled and unfilled sample is not influenced substantially by the processing routes. The decreased width at half height in the first heating curves (W_h^{1st}) of blown films indicates a narrow distribution of crystallite sizes, relative to the compression molded samples, due to rapid cooling and more efficient nucleation efficiency of disentangled nanotubes.

4.4.3 Tensile Properties

Tensile tests are carried out to investigate the mechanical properties of neat HDPE and HDPE/MWCNT composite films containing 4 wt% MWCNTs. Figure 4.8 shows the tensile strain-stress curves of samples at BURs 2 and 3. The Young's modulus, stress at break, and strain at break of blown films with increasing BURs are shown in Figure 4.9 and compared with compression molded samples. The experimental modulus is compared with the theoretical values from a Halpin–Tsai composite model. According to the Halpin–Tsai model, the longitudinal (E_{11}) and transverse (E_{22}) moduli can be predicted using Eqs. (4.1) and (4.2), respectively. However, for the compression molded composite sheets, the Halpin–Tsai model can be expressed as Eq. (4.3).

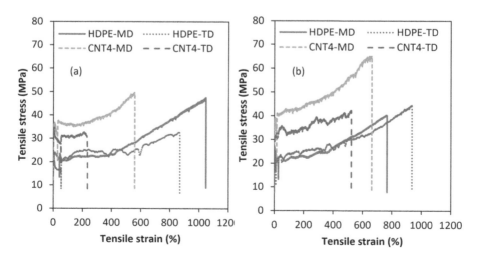

FIGURE 4.8 Tensile strain-stress curves of neat HDPE and HDPE/MWCNT composite blown films containing 4 wt% MWCNTs at BURs of 2 (a) and 3 (b).

Blown Film Extrusion of Conductive Polymer Composites

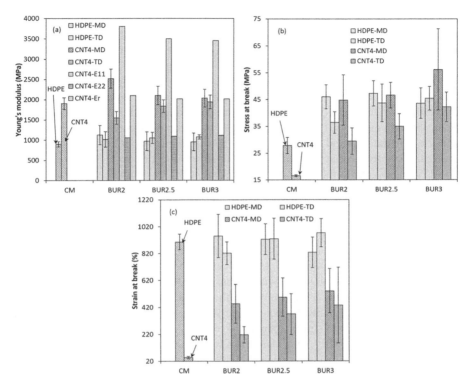

FIGURE 4.9 Young's modulus (a), stress at break (b), and strain at break (c) of the blown films of neat HDPE and HDPE/MWCNT (4 wt%) composite with increasing BURs. (In Panel (a), the modulus of the matrix used for the calculation of CNT4-E_r is the average modulus of HDPE film in the MD and TD at each BUR).

$$\frac{E_{11}}{E_m} = \frac{1+2\alpha\eta_L\varnothing_f}{1-\eta_L\varnothing_f} \qquad (4.1)$$

$$\frac{E_{22}}{E_m} = \frac{1+2\eta_T\varnothing_f}{1-\eta_T\varnothing_f} \qquad (4.2)$$

$$\frac{E_r}{E_m} = \frac{3}{8}\left[\frac{1+2\alpha\eta_L\varnothing_f}{1-\eta_L\varnothing_f}\right]+\frac{5}{8}\left[\frac{1+2\eta_T\varnothing_f}{1-\eta_T\varnothing_f}\right] \qquad (4.3)$$

where $\eta_T = (E_f/E_m - 1)/(E_f/E_m + 2)$ and $\eta_L = (E_f/E_m - 1)/(E_f/E_m + 2\alpha)$. E, E_m, and E_f are the elastic modulus of composites, matrix, and filler, respectively. E_f is set as 200 GPa in this case, $\varnothing_f$ is the volume fraction of filler, α is the aspect ratio of the filler, set as 150 according to the MWCNT dimensional parameters.

Figure 4.8 indicates that the blown film of neat HDPE exhibits a higher tensile strength in the MD than TD at a BUR of 2, while this phenomenon is reversed at a

BUR of 3. This may be attributed to some of the lamellae stacked perpendicular to the MD, changing their orientation to the TD, with increasing BURs [19]. The tensile stress for the HDPE/MWCNT composite films is higher than that of the neat HDPE films at the same strain, associated with the reinforcing effect of the MWCNTs [20], while the elongation of the composite films decreases markedly compared with the neat HDPE. In addition, the higher tensile strength always occurs in the MD for the blown films of the composite, indicating that more MWCNTs are probably aligned along the MD even at a BUR of 3. This may be due to less potential for molecular relaxation in the MD and more rapid cooling, resulting in better heat transfer [21].

Figure 4.9a shows that the modulus of the blown films of neat HDPE increases by 6%–25% on average in the MD and TD at different BURs, compared to the compression molded HDPE sample, this may be associated with the orientation of polymer chains. An evident anisotropy in modulus can be observed for the composite with 4 wt% MWCNTs at a BUR of 2, where the modulus in the MD increases by 32%; however, in TD, it decreases by 19% compared to the compression molded sample. At this BUR, the higher modulus is attributed to better MWCNTs orientation along the MD. Then the difference in modulus in the MD and TD decreases gradually since some MWCNTs are reoriented along TD with increasing BURs (Figure 4.6a and b). At BURs of 2.5 and 3, the modulus of the composite thin films does not improve significantly, in both MD and TD, relative to the compression molded sample. Figure 4.9b and c reveal that the stress and strain at the break of the blown films of HDPE/MWCNT composite increase steadily with increasing BURs, due to the breakup of MWCNT agglomerates. The strain at break of neat HDPE is not influenced by blown film extrusion (Figure 4.9c), while the stress of all the blown films of neat HDPE increases by about 57%, mainly attributed to the enhanced orientation of polymer chains; however, the crystallinity of neat HDPE is not increased, as shown in Table 4.5.

4.4.4 Electrical Properties

The change in volume resistivity of blown films of HDPE/MWCNT composites with different MWCNT loadings as a function of BUR is investigated, and the results are shown in Figure 4.10.

As expected, the resistivity of neat HDPE is barely influenced by BUR (Figure 4.10a). However, the resistivity of the composite with 2 wt% MWCNTs increases by 10 orders of magnitude at a BUR of 2 before leveling off. At this BUR, the distance between the MWCNTs exceeds the critical maximum distance of 1.8 nm for electron hopping [22–25]. For the composite with 4 wt% MWCNTs, the resistivity of blown films increases steadily with increasing BUR, and it also increases 10 times when the BUR increases to 3. However, there is no significant increase in the resistivity of the composite containing 8 wt% MWCNTs after film blowing at higher BUR. This implies that at this loading, a sufficient density of nanotubes forms a robust conductive network in the matrix. Figure 4.10b shows that the resistivity of the composite film with 8 wt% MWCNTs is very close in both MD and TD; however, the resistivity in the MD is slightly lower than that in the TD, probably due to better alignment of MWCNTs [26–28].

To investigate the effect of annealing on the volume resistivity of the composite films, composites containing 4 wt% MWCNTs are annealed under different

Blown Film Extrusion of Conductive Polymer Composites

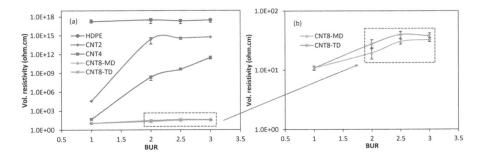

FIGURE 4.10 Changes in the volume resistivity of the blown films of HDPE/MWCNT composites with increasing BURs (the BUR of 1 refers to the compression molded samples): (a) overall view for the composites containing different MWCNT loadings, (b) enlarged view for CNT8-MD and CNT8-TD.

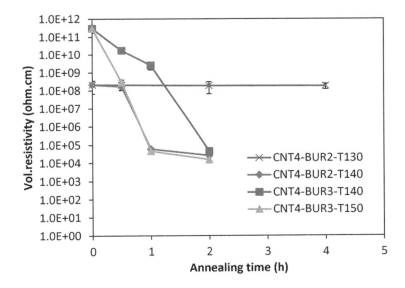

FIGURE 4.11 Changes in the volume resistivity of HDPE/MWCNT composite films with 4 wt% MWCNTs after annealing.

conditions. The changes in the volume resistivity of blown films of HDPE/MWCNT composite with 4 wt% MWCNTs after annealing at 140°C for 2 hours are shown in Figure 4.11. The resistivity of composite film at a BUR of 2 is barely influenced by annealing at 130°C; however, it decreases by four orders of magnitude at 140°C for 1 hour. Similarly, the resistivity of composite films at a BUR of 3 decreases by 7 orders of magnitude after annealing at 140°C for 2 hours or annealing at 150°C for 1 hour. A higher annealing temperature benefits the recovery of the MWCNTs conductive network, attributed to the decreased viscosity of the polymer matrix, and increased mobility of nanotubes at a higher annealing temperature, thereby facilitating the reformation of local contacts between MWCNTs and reaggregation [29]. Composite thin films with a smaller BUR (less deformation) require a shorter time to repair the conductive network during annealing. Some previous studies [17,29] have also found that the electrical properties of polymer/MWCNT composites after extensional deformation are enhanced by annealing at an appropriate temperature.

4.5 BLOWN BUBBLE FILMS OF ALIGNED NANOWIRES AND CNTs

With the continuous expansion of the technical fields, nanowires and nanotubes are directionally assembled on a large area in many practical applications. Although directional assembly and Langmuir–Blodgett methods have made significant progress, it is still unclear whether these technologies can be extended to large wafers and non-rigid substrates. Yu et al. [30] used blown film extrusion to manufacture nanocomposite films and described a general and scalable method for large-area, uniformly arranged, and density-controlled nanowire and nanotube films. This method can transfer the blown bubble film (BBF) to a single crystal wafer with a diameter of 150~200 mm and a size greater than 225×300 mm^2 flexible plastic sheet and highly curved surface. This type of wafer can be hung on an open frame.

4.5.1 BBFs WITH SILICON NANOWIRES

Silicon nanowires are covalently modified with 5,6-epoxyhexyl triethoxysilane and combined with an epoxy solution to produce a stable and well-dispersed suspension (Figure 4.12a). The nitrogen flow blows in a single bubble and generates stable vertical expansion bubbles under the upward-moving ring. Figure 4.12b reveals that two 150 mm silicon wafers are fixed near the mold and the central axis of the bubble. The bubble continues to expand until it covers the entire crystal surface, as shown in Figure 4.12c. The optical detection activity is performed uniformly on the surface of the wafer with a diameter of 150 mm (Figure 4.12d), which shows that the transfer film is uniform on the whole substrate. The high magnification dark-field optical image analyzes a single silicon nanowire in the transmission BBF, indicating that the silicon nanowires recorded from different areas of the large substrate are well arranged along the upward expansion direction of the bubble. The angular deviation of silicon nanowires is <10° on the whole wafer with a diameter of 150 mm, suggesting great progress over the previous research. In addition, the BBF method can also transfer the oriented Si NW BBF film to a wide range of substrates. The Si NW BBF is transferred to the semicircular column, and the subsequent dark-field optical images confirm that the NW in the blown film is well aligned, as shown in Figure 4.12e [31]. In addition to flat and curved substrates, Si NW BBF is also transferred to open frames (Figure 4.12f), verifying the great flexibility of this method. Importantly, the BBF method also has the potential to be extended to ultra-large area structures. Si NW-BBF can be evenly transferred to the large rectangular plastic plate substrate of 225×300 mm (Figure 4.12g) with controlled arrangement and density of Si NW. A histogram of angle distribution of over 400 Si NWs taken from different locations over the entire plastic substrate shows that more than 85% of NWs are aligned within ±6° of the primary expansion/alignment direction.

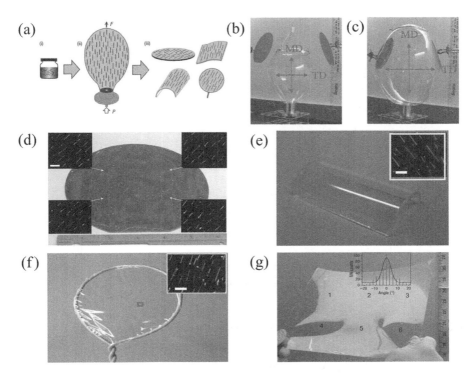

FIGURE 4.12 Blown bubble film process. (a) Schematic diagram of NW/CNT polymer suspension blown film extrusion. (b and c) Photographs of directed bubble expansion process at early and final stages, respectively. (d) NW-BBF dark-field optical image transferred to the silicon wafer. (e) Image of a 0.10 wt% Si NW BBF transferred to a curved surface. (f) Image of a 0.10 wt% Si NW BBF transferred to an open frame with a diameter of ~6 cm. (g) Image of a 0.10 wt% Si NW BBF transferred to a 225 mm × 300 mm plastic substrate.

Figure 4.13 suggests that with the increase in concentration, the separation degree of Si NW decreases significantly, and the density of the transferred BBF increases. The NW separation can be varied over at least an order of magnitude from 50 ± 8 to $3.0 \pm 0.6\,\mu m$ as concentration increases from 0.01 to 0.22 wt%. Correspondingly, NW density increases from $4.0 \pm 0.6 \times 10^4$ to $4.0 \pm 0.5 \times 10^6\,cm^{-2}$ for these same samples. So far, the separations/densities of Si NWs BBF are relatively small; however, it is still suitable for some applications, such as nano-electronic transistor arrays for biological/chemical sensing and displays. At higher Si NW concentrations, the saturation is close to the micron spacing, attributed to the observed NW aggregation to some extent.

4.5.2 BBFs with CNTs

Due to the versatility of this method in nanowire and nanotube materials, substrate structure and size scaling are further explored. Single-walled carbon nanotubes (SWCNTs) and MWCNT are modified with *n*-Octadecylamine, and the resulting suspension is used to prepare BBFs. The transferred SWCNT-BBF and MWCNT-BBF

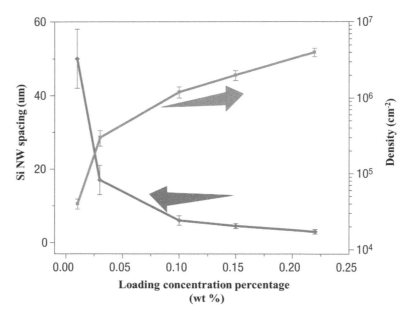

FIGURE 4.13 Plot of the average Si NW spacing and density as a function of Si NW loading.

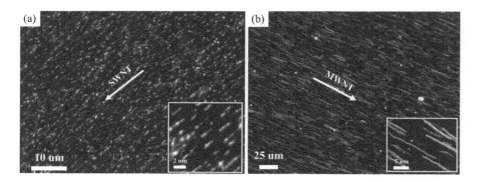

FIGURE 4.14 Dark-field optical image of (a) SWCNT-BBF prepared from 0.07 wt% solution and (b) MWCNT-BBF prepared from 0.15 wt% solution.

show good alignment and uniformity on the 75 mm diameter substrate and can be transferred to a larger substrate (Figure 4.14a and b). The average spacing of SWCNTs is $1.5 \pm 0.4\,\mu m$, and 90% of SWCNTs are arranged within 5° of the average orientation. The good alignment of SWCNT in BBF is noteworthy because of their length, ~1 to $2\,\mu m$, about ten times that of MWCNTs (~20 to $25\,\mu m$) and Si NWs (~10 to $15\,\mu m$). It was also found that longer MWCNT, which initially curled a little, straightened in BBF.

In addition, CNT-BBF films can also be transferred to a wide range of substrates, such as planar, curved substrates, and open frames. The CNT-BBF film is uniformly

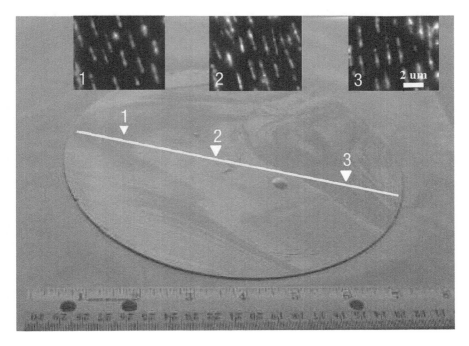

FIGURE 4.15 Image of 0.07 wt% SWCNT-BBF transferred to a 200 mm Si wafer and its line-scanning alignment analysis.

transferred to the 200 mm wafer (Figure 4.15). The dark-field optical image shows that SWCNT has the same direction and uniform separation on this large substrate.

4.5.3 BBFs with Large-Area Transistor Arrays

The high alignment, controlled density, and large area coverage of NW-BBF and CNT-BBF may open new pathways in electronic applications of these nanomaterials possible. Figure 4.16a shows that Si NW BBF is directly transferred to a diameter 75 mm plastic substrate, and an independently addressable NW-FET (field-effect transistor) array is fabricated.

Representative drain-source current (I_{ds}) versus gate voltage (V_g) yield a peak transconductance, $g_m = dI_d/dV_g$, of 6 μS with an on current (Ion) of ~16 μA, an on/off ratio >105, and a threshold voltage (V_t) of 0.55 V (Figure 4.16b). These values are comparable to or exceed, previous multi-Si NW FETs prepared using Langmuir–Blodgett assembly, and significant improvements are possible, i.e., by substituting much higher performance Ge/Si core/shell NWs. Importantly, histograms of V_t and Ion (Figure 4.16c) and inset of Figure 4.16b), show that these properties, critical to integrated systems, are well constrained, with values of 0.81 ± 0.32 V and 15.1 ± 3.7 μA, respectively. The good reproducibility of the Si NW FETs is attributed to the uniform density, good alignment, and preferential distribution of the NWs at a single surface of BBFs, as this allows the fabrication of repeatable device structures. The

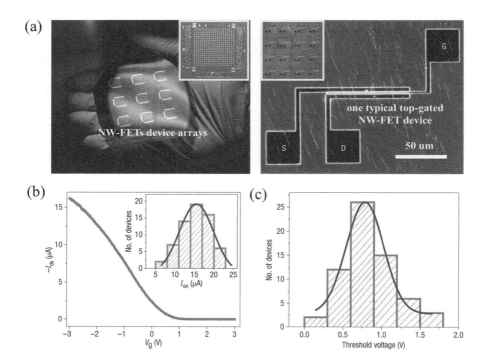

FIGURE 4.16 (a) Si NW FET arrays on plastic substrates. (b) Typical I_{ds}–V_g characteristics of a 12-Si NW FET device recorded with $V_{ds} = -1$ V. Inset, histogram of I_{on} showing the uniform device characteristics, where the blue curve is a Gaussian fit: 15.1 ± 3.7 μA. (c) Histogram of threshold voltage determined from analysis of more than 60 randomly chosen devices in the larger array; the blue curve is a Gaussian fit: 0.81 ± 0.32 V.

straightforward transfer of aligned Si NW BBFs to large substrates makes this process considerably more efficient than previous fluid-directed and Langmuir–Blodgett assembly methods.

4.6 CONCLUSIONS

The film blowing extrusion process is closely related to the properties of raw materials. For different types of plastics, the process of blown film extrusion is also different. At present, LDPE, HDPE, LLDPE, EVA, and ion-bonded polymers are mainly used in blown film production. EVA and ion-bonded polymers are commonly employed in coextrusion. Due to the different MFR (melt flow rate) value and molecular weight distribution width of these materials, they have their characteristics in the selection of equipment and process conditions.

In this chapter, the effects of blown film extrusion process on the structure and properties of TPU/MWCNT, TPU/GO, and HDPE/MWCNT composite films are reviewed. Due to the special structure, TPU/MWCNT composites suggest that the higher the content of MWCNT, the larger the content and size of segregated hard domains. The blown film extrusion process allows better self-assembly of hard domains and maximizes the distance between domains. For blown film HDPE/MWCNT composites, the addition of MWCNTs improves the elastic response during deformation, resulting in a more uniform thickness distribution. Due to the enhanced orientation and dissociation of MWCNTs, blown film composites show better mechanical properties. The resistivity of the composite film increases significantly with the increase in blow-up ratio. These paths can be partially restored using an appropriate annealing process.

In addition, blown film extrusion is a scalable method for large-area, uniformly arranged, and density-controlled nanowire and nanotube films. The BBF can be transferred to a single crystal wafer with a diameter of at least 200 mm, a flexible plastic sheet with a size of at least 225×300 mm^2, and a highly curved surface.

REFERENCES

1. Karkhanis SS, Stark NM, Sabo RC, Matuana LM. Blown film extrusion of poly(lactic acid) without melt strength enhancers. *J Appl Polym Sci.* 2017;34(34):0021–8995.
2. Shirai MA, Olivato JB, Garcia PS. Thermoplastic starch/polyester films: effects of extrusion process and poly (lactic acid) addition. *Mater Sci Eng C-Mater Biol Appl.* 2013;33(7):4112–7.
3. Li S, Luo JX, Gao L, Mao LX. Study on processing and property of three-layer co-extrusion barrier films. *Zhongguo Suliao.* 2011;25(10):19–23.
4. Chang XD. Study on self reinforcing technology and mechanism of polyethylene film blow molding process. South China University of Technology. 2020.
5. Ge C, Aldi R. Effects of aragonite calcium carbonate on barrier and mechanical properties of a three-layer co-extruded blown low-density polyethylene film. *J Plast Film Sheeting.* 2014;30(1):77–90.
6. Zhu J, Gao W, Wang B, Kang XM, Li PF, et al. Preparation and evaluation of starch-based extrusion-blown nanocomposite films incorporated with nano-ZnO and nano-SiO$_2$. *Int J Biol Macromol.* 2021;183:1372–8.
7. Guichon O, Seguela R, David L, Vigier G. Influence of the molecular architecture of low-density polyethylene on the texture and mechanical properties of blown films. *J Polym Sci Part B (Polym Phys).* 2003;41(4):327–40.
8. Fatahi S, Ajji A, Lafleur PG. Correlation between different microstructural parameters and tensile modulus of various polyethylene blown films. *Polym Eng Sci.* 2010;47(9):1430–40.
9. Antunes M, Mudarra M, and Velasco JI. Broad-band electrical conductivity of carbon nanofibre-reinforced polypropylene foams. *Carbon.* 2011;49(2):708–17.
10. Russo P, Lavorgna M, Piscitelli F, Acierno D, Dimaio L. Thermoplastic polyurethane films reinforced with carbon nanotubes: The effect of processing on the structure and mechanical properties. *Eur Polym J.* 2013;49(2):379–88.
11. Xia HS, Song M. Preparation and characterization of polyurethane–carbon nanotube composites. *Soft Matter.* 2005;1(5):386–94.
12. Russo P, Acierno D, Capezzuto F, Buonocore G, Dimaio L, et al. Thermoplastic polyurethane/graphene nanocomposites: The effect of graphene oxide on physical properties. *AIP Conf Proc.* 2015; 1695: 020030.
13. Xiang D, Harkin-Jones E, Linton D, Martin P. Structure, mechanical, and electrical properties of high-density polyethylene/multi-walled carbon nanotube composites processed by compression molding and blown film extrusion. *J Appl Polym Sci.* 2015;132(42):1–12.

14. Khan SU, Pothnis JR, Kim JK. Effects of carbon nanotube alignment on electrical and mechanical properties of epoxy nanocomposites. *Compos Part A Appl Sci Manuf.* 2013;49:26.
15. Alig I, Skipa T, Engel M, Lellinger D, Pegel S, Potschke P. Electrical conductivity recovery in carbon nanotube-polymer composites after transient shear. *Phys Status Solidi B.* 2007;244(11):4223–6.
16. Jiang Z, Chen YH, and Liu Z. The morphology, crystallization and conductive performance of a polyoxymethylene/carbon nanotube nanocomposite prepared under micro-injection molding conditions. *J Polym Res.* 2014;21(6):1–15.
17. Mayoral B, Hornsby PR, Mcnally T, Schiller TL, Jack K, et al. Quasi-solid state uniaxial and biaxial deformation of PET/MWCNT composites: Structural evolution, electrical and mechanical properties. *RSC Adv.* 2013;3(15): 5162–83.
18. Shen JB, Champagne MF, Yang Z, Yu Q, Gendron R et al. The development of a conductive carbon nanotube (CNT) network in CNT/polypropylene composite films during biaxial stretching. *Compos Part A Appl Sci Manuf.* 2012;43(9):1448–53.
19. Zhang J, Liu FH, Qian XY, Lei YW. Studies on preparation of HDPE/CB composites including a novel oriented structure by the microwave heating and their characterization. *Polym Adv Technol.* 2011;22(6):811–6.
20. Istrate OM, Paton KR, Khan U, Neill AO, Bell AP, Coleman JN. Reinforcement in melt-processed polymer-graphene composites at extremely low graphene loading level. *Carbon.* 2014;78:243–9.
21. Han ZD, Fina A. Thermal conductivity of carbon nanotubes and their polymer nanocomposites: A review. *Prog Polym Sci.* 2011;36(7):914–44.
22. Antunes RA, Oliveira MCL, Ett G. Carbon materials in composite bipolar plates for polymer electrolyte membrane fuel cells: A review of the main challenges to improve electrical performance. *J Power Sources.* 2011;196(6):2945–61.
23. Sengupta R, Bhattacharya M, Bandyopadhyay S. A review on the mechanical and electrical properties of graphite and modified graphite reinforced polymer composites. *Prog Polym Sci.* 2011;36(5):638–70.
24. Yu Y, Song G, Sun L. Determinant role of tunneling resistance in electrical conductivity of polymer composites reinforced by well dispersed carbon nanotubes. *J Appl Phys.* 2010;108(4):084319.
25. Brkovi DV, Kovacevic VV, Sretenovic GB, Kuraica MM, Trisovic NP, et al. Effects of dielectric barrier discharge in air on morphological and electrical properties of graphene nanoplatelets and multi-walled carbon nanotubes. *J Phys Chem Solids.* 2014;75(7):858–68.
26. Wen M, Sun XJ, Su L, Shen JB, Li J, et al. The electrical conductivity of carbon nanotube/carbon black/polypropylene composites prepared through multistage stretching extrusion. *Polymer.* 2012;53(7):1602–10.
27. Al-Saleh MH, Gelves GA, Sundararaj U. Copper nanowire/polystyrene nanocomposites: Lower percolation threshold and higher EMI shielding. *Compos Part A Appl Sci Manuf.* 2011;42(1):92–7.
28. Jeon K, Lumata L, Tokumoto T, Steven E, Brooks James, et al. Low electrical conductivity threshold and crystalline morphology of single-walled carbon nanotubes: High density polyethylene nanocomposites characterized by SEM, Raman spectroscopy and AFM. *Polymer.* 2007;48(16):4751–64.
29. Zhang SM, Lin L, Deng H, Gao X, Bilitti E, et al. Dynamic percolation in highly oriented conductive networks formed with different carbon nanofillers. *Colloid Polym Sci.* 2012;290(14):1393–1401.
30. Yu GH, Cao A, Lieber CM. Large-area blown bubble films of aligned nanowires and carbon nanotubes. *Nat Nanotechnol.* 2007;2(6):372–7.
31. Cantor K. Blown film extrusion: An introduction. *Plast Technol.* 2007;53(1):34.

5 Temperature-Resistivity and Damage Self-Sensing Behavior of Conductive Polymer Composites

5.1 INTRODUCTION

Conductive polymer composites (CPCs) are multiphase composites made by inserting conductive fillers into polymers and compounding them via physical or chemical methods [1–4]. Compared with traditional conductive metal materials, polymer-based conductive composites have the characteristics of lightweight, easy processing into various complex shapes, corrosion resistance, and resistivity adjustable in a large range. CPCs can be used as electromagnetic shielding materials, bipolar plates for fuel cells, self-limiting heating composites, overcurrent protectors, and so on, having broad market prospects in many fields of national economy and national defense industry [5].

Positive temperature coefficient (PTC) is a common temperature-resistivity parameter according to which the electrical resistivity of CPCs increases with the rising temperature. However, when the temperature is beyond the melting point (T_m) of the polymer matrix, carbon black (CB)-filled non-cross-linking semi-crystalline polymer composites always exhibit an obvious decrease in their resistivity, and this phenomenon is attributed to the negative temperature coefficient (NTC) [6]. Therefore, the temperature-resistivity behavior of CPCs can mirror the variation of their conductive networks with temperature. In order to process CPCs into a desired final product, they usually experience some phase transition processes such as melting and crystallization with the temperature profile. The deformation or reorganization of conductive networks in polymers during such processing could significantly alter the electrical conductivity of the final product compared to that of the original unprocessed material. In this chapter, attention will be paid to the approach for in situ characterization of the electrical conductivity of CPCs with randomly distributed structure and segregated network structure during melting and crystallization, which combines differential scanning calorimetry (DSC) and electrical conductivity measurements. In addition, on the basis of temperature-resistivity behavior, the temperature sensor as a kind of CPC application will be also briefly reviewed.

CPCs represent excellent electrical conductivity and considerable mechanical strength, which are used to develop smart structural materials, including damage self-sensing (or damage self-monitoring) polymer composites that are able to provide structural and sensing properties by themselves [7–9]. In this chapter, the damage

DOI: 10.1201/9781003218661-5

self-sensing behavior of the high-density polyethylene (HDPE)-based CPCs investigated via in situ electromechanical measurements is presented. It is shown that nanofiller types and conductive network structures, as well as excessive nanofiller loading, significantly influence the damage self-sensing behavior of the composites. Besides, the above technique is also suitable for structural thermosetting materials [10,11].

5.2 CONSTRUCTION OF CONDUCTIVE NETWORK STRUCTURES IN CPCs

As most ordinary polymeric matrices are insulating, the conductivity of CPCs relies on the content of incorporated conductive fillers. When the amount of such fillers reaches a critical value, CPCs will exhibit an insulator/conductor transition, during which their electrical conductivity dramatically increases by several orders of magnitude [12,13]. This insulator/conductor transition is thought to be associated with the formation of conductive paths throughout polymeric matrices by directly linked conductive fillers or by the tunneling current between nearby conductive fillers where electrons can flow through the insulating barrier via quantum mechanical tunneling between adjacent conductive regions [14]. The conductivity of CPCs can be quantitatively analyzed using a scaling law of percolation threshold as follows [15]:

$$\sigma_{DC} \propto (\varnothing - \varnothing_c)^t \tag{5.1}$$

where σ_{DC} is the electrical conductivity of CPCs, $\varnothing$ is the filler weight fraction, $\varnothing_c$ is the filler weight fraction at the percolation threshold, t represents the critical exponent depending on the dimensionality of the conductive network [16,17]. It can be seen that the coefficient t, corresponding to the conductive network structure, poses a significant influence on the conductivity of CPCs [18]. Therefore, the conductivity of CPCs can be tailored by constructing appropriate conductive networks.

According to the conductive filler distribution in the matrices, most CPCs can be categorized into the following three kinds with different conductive network structures: CPCs with randomly dispersed structure (r-CPCs), CPCs with segregated structure (s-CPCs), and CPCs with double percolation structure (d-CPCs). On the basis of the "process/structure/property" relationship, the structure of a conductive network corresponds to the preparation method of CPCs, which is specifically associated with types both of host polymer and conductive fillers, as well as process routes.

Conductive fillers randomly dispersed in polymeric matrices are beneficial for CPCs. These can be achieved using some ordinary preparing routes such as melt blending and solution mixing [19–21]. Melt blending is quite an appropriate choice for preparing r-CPCs on an industrial scale, which is because of its facile operation and massive output [22]. Villmow et al. [23] systematically studied the effect of extrusion parameters on the dispersion of conductive fillers in CNT/PCL composites. It was found that an increase in the rotation speed significantly decreased the CNT agglomerate size in PCL and improved their dispersion. The higher shear stress arising from the higher rotation speeds was more likely to overcome cohesive strengths and physical entanglements of CNT agglomerates. In addition, one can observed that higher residence time corresponded to a better dispersion of CNT, and the rotation speed and the screw configuration are two main factors to influence the residence time. Although the rotation speed is inversely related to the residence time, choosing appropriate screw configurations such as using back-conveying elements and extending screw length will increase the residence time, and therefore improves melt-blending effect and gives arise to better dispersion of CNT.

Solution mixing is another technique to prepare r-CPCs, which consists in dissolving thermoplastic polymer granules into a suitable solvent, mixing the obtained blend with conductive fillers, and evaporating the solvent. This method allows fillers to mix with dispersed polymer chains, leading to a locally homogenous dispersion state [12,24]. Concerning thermosetting matrices, an example using solution mixing to fabricate r-CPCs was reported by Ajayan et al. [25]. In that work, multi-walled carbon nanotubes (MWCNTs) were firstly dispersed in ethanol by sonication and mechanically stirred with a mixture of epoxy monomer and curing agent. After evaporation of the solvent, the MWCNTs-epoxy mixture was poured into molds and cured to achieve solids.

As for the s-CPCs and d-CPCs materials, their conductive fillers are both purposively distributed in the matrices. In the former system, fillers are primarily located at the interfaces between the polymeric matrix particles, and the "double percolation" indicates conductive fillers are selectively located in one phase of a co-continuous polymer blend. Due to these special distributions, lower contents of conductive fillers are needed to achieve the insulator/conductor transition, which results in the reduction in the percolation value (Pc) several times compared to conventional r-CPCs. For instance, Xiang et al. [26] prepared MWCNTs-filled HDPE CPCs with randomly dispersed conductive network (DCN) and segregated conductive network (SCN) structures. It was found that the Pc value of CPCs in the SCN structure (0.1 wt%) was lower than that of CPCs in the DCN structure (0.5 wt%), which indicated the segregated conductive network constructed more effective conductive paths in CPCs.

Hot compressing is a common process to prepare s-CPCs, prior to which a mixture of polymer granules with conductive fillers is first obtained via dry blending or solvent dispersion. As the polymer granules are dispersed in an inactive solvent rather than being dissolved, a mixture with conductive fillers-coated polymer granules is obtained after evaporating the solvent. During hot compressing, the melted polymer with high viscosity reduces the mobility of fillers, resulting in a large amount of fillers localized in the boundary of polymer phases [27]. Besides, melt blending is often used to fabricate s-CPCs with a selective distribution of conductive fillers at the interfaces of immiscible polymer compounds [28]. Gubbels et al. [29] fabricated s-CPCs with CBs distributed through the immiscible polyethylene/polystyrene (PE/PS) blend interface. Particularly, PS (or PE) and CB were first stirred in an internal mixer. After the PS/CB or PE/CB mixture was effectively melt blended, the second polymer, i.e. PE (or PS), was then added. On the basis of thermodynamic drive, CBs would gradually migrate from one polymer phase to the other. Therefore, the above s-CPCs can be obtained via controllable migration of CBs.

As for d-CPCs materials, Zhang et al. [30] produced the PP/PE/CNTs-based CPCs with a double-percolated conductive network. In the preparation, CNTs were first dispersed in PE and the obtained PE/CNTs composites were then compounded with PP particles well below the T_m value of PP. Pang et al. [31] also reported CNT/

HDPE/ultra-high density polyethylene (UHMWPE) composites with segregated and double-percolated network structures, which were successfully obtained using high-speed mechanical mixing and hot compression. According to the results, the d-CPCs system consisted of at least two polymer phases, and in one of them which conductive fillers selectively located constructed continuous conductive paths throughout CPCs. Therefore, the phase morphology of the immiscible blends plays an important role in the electrical properties of d-CPCs. In another work, Xu et al. [32] studied the effect of PLA and PCL ratios on the conductivity of d-CPCs filled with MWCNTs. It was found that the conductivity of PLA/PCL/MWCNT composites (1 wt% MWCNTs) increased by increasing PCL content, achieving its maximum value at the PCL concentration of 40–50 wt%. In addition, several groups [33,34] have shown that the choice of polymer matrices can significantly influence the distribution of conductive fillers in the polymer blends. This is because of different interactions between polymers and fillers, as well as the location of conductive fillers in the polymer phase they more interact with.

5.3 STIMULI-RESISTIVITY BEHAVIORS OF CPCs

As discussed above, s-CPCs and d-CPCs with specifically distributed conductive fillers both exhibit the lower P_c value or the earlier insulator/conductor transition than that of r-CPCs, which is owing to plenty of effective conductive networks therein. It is widely known that conductive networks are transformed under the action of some external stimuli such as temperature fluctuation and applied stress, which results in the corresponding variation in the conductivity of CPCs. In this section, we would like to discuss the influence of temperature and applied strain on the resistivity (1/conductivity) of CPCs with different conductive networks.

5.3.1 Temperature-Resistivity Behavior

PTC and NTC effects are often observed in CPCs, and they are respectively indicatives of an increase and decrease in their resistivity with the increasing temperature [35]. Fundamentally speaking, such a characteristic of temperature-resistivity behavior takes its origin from the structural transformation of a conductive network in CPC. Xiang [36] prepared HDPE/MWCNT and HDPE/graphene nanoplatelet (GNP)-based r-CPCs and segregated structure, respectively, via solution-assisted mixing. Their temperature-resistivity behaviors were then studied by in situ characterization combining the DSC with electrical conductivity measurements. The electrical and thermal properties of some representative specimens are shown in Figure 5.1a and b, and the temperature profile used in the experiments is illustrated in Figure 5.1c. Based on the melting temperature (T_m) and crystallization temperature (T_c) of the specimens listed in Table 5.1, it is clear that temperature-resistivity behaviors of all samples underwent the entire melting and crystallization processes.

As shown in Figure 5.1a and b, the first PTC effect is observed prior to the onset of melting, by which the resistance change ratios ($\Delta R/R_0$) of the HDPE/MWCNT and HDPE/GNP composites increase gradually. This is mainly because of the volume expansion of the polymer matrix, resulting in the inferior connection between nearby MWCNTs. In addition, the increase in $\Delta R/R_0$ for the CD-1 (r-CPCs with

Temperature-Resistivity and Damage Self-Sensing Behavior

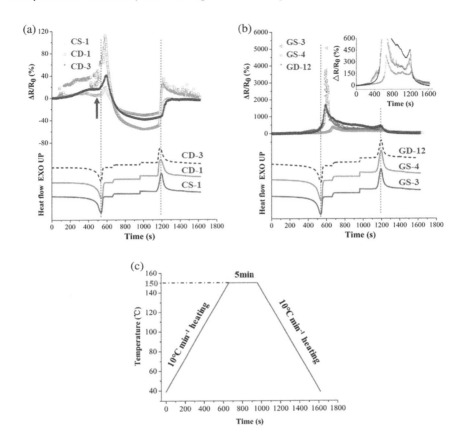

FIGURE 5.1 In situ electrical measurement results for (a) HDPE/MWCNT and (b) HDPE/GNP composites during phase transitions and (c) temperature profile used in the experiments.

TABLE 5.1
Summary of Parameters for Nanocomposites from DSC and XRD Analyses

Sample	Nanofiller Content (wt%)	T_m (°C)	T_c (°C)	X_{DSC} (%)	X_{XRD} (%)	L_{110} (nm)	L_{200} (nm)
HDPE	0	129.6	108.2	43.0	54.5	20.1	16.9
CS-1	1	128.2	111.8	44.5	56.9	21.2	17.8
CD-1	1	128.5	113.2	45.3	56.8	20.4	17.5
CD-3	3	128.3	114.5	46.2	61.6	21.1	18.0
GS-3	3	128.6	111.9	44.0	56.5	20.4	16.4
GS-4	4	129.0	111.9	45.9	59.9	20.7	17.0
GD-12	12	128.9	112.2	42.3	48.0	21.7	18.0

HDPE, high-density polyethylene; DSC, differential scanning calorimetry; XRD, X-ray diffraction; CS, CNT composite with a segregated structure; CD, CNT composite with a dispersed structure; GS, GNP composite with a segregated structure; GD, GNP composite with a dispersed structure.

1 wt% MWCNT) is larger compared to that of CS-1 (s-CPCs with 1 wt% MWCNT) and CD-3, indicating there was a less robust conductive network in the former composite. A slight decrease in $\Delta R/R_0$ for the HDPE/MWCNT composites during the onset and peak of melting (marked by an arrow in Figure 5.1a) can be attributed to the motion of nanotubes to regenerate some interconnected contacts [37], while this NTC effect is absent in the HDPE/GNP composites probably due to the lower mobility of the two-dimensional GNPs. Subsequently, the values of $\Delta R/R_0$ for both MWCNT- and GNP-filled composites drastically increased up to the end of the melting stage. This second PTC effect stemmed from the transformation of the crystalline phase into the amorphous state, which was accompanied by a significant volume expansion, resulting in an increase in the distance between the conductive fillers and the reduction in the probability of electron tunneling between nearby conductive regions [38]. Compared with HDPE/MWCNT composites, a higher increase in $\Delta R/R_0$ was observed in HDPE/GNP composites, which indicated that the conductive network formed by GNP was more readily damaged due to the lack of entanglements or interlacing compared to that in the one-dimensional nanotubes. The second NTC effect occurred after the completion of melting, during which the $\Delta R/R_0$ values significantly decreased for the HDPE/MWCNT and HDPE/GNP composites and leveled off until the onset of crystallization. This is because HDPE with enough low viscosity after complete melting accelerated the motion of conductive fillers in the matrix and promoted the reorganization of the conductive network.

One can see that repeated PTC and NTC effects emerged in the above specimens with the increasing temperature. The PTC effect is thought to be related to the volume expansion of the matrix, induced by the volume expansion of a solid matrix or by the melting behavior. In contrast, the NTC effect is associated with the reconstruction of effective conductive networks. In addition, these temperature-resistivity performances are significantly affected by the original network structure in the polymer as well as the shape and loading of nanofillers.

Because of their temperature-resistivity behavior, the CPC materials find application in temperature sensitive devices such as temperature sensors [39–41]. A desirable temperature sensor usually requires a low I_{NTC} value to maintain highly responsive signal toward the temperature stimuli [42]. As discussed above, the NTC effect is associated with the reconstruction of conductive networks, and the s-CPC materials have a more robust conductive network. Therefore, constructing a segregated conductive network in the temperature sensor is expected to eliminate the NTC effect. For instance, a CB/UHMWPE s-CPC-based temperature sensor without the NTC effect was prepared by Zhang et al. [43]. Since CB was concentrated at the interspace of UHMWPE granules and formed a relatively intensive conductive network, there was little room for fillers to reconstruct more effective conductive paths with the increasing temperature. In addition, the ultra-high melt viscosity of the UHMWPE matrix also hindered CB particles from reconstructing the conductive network.

High I_{PTC} is also desirable for temperature sensors because the obvious response to the temperature fluctuation is required. Therefore, it is important to tune I_{PTC} of CPCs. Although by decreasing the content of conductive fillers to construct a less stable conductive network in temperature sensors can increase their I_{PTC}, the lower responds current results in more obvious fluctuation on its feedback signals.

Temperature-Resistivity and Damage Self-Sensing Behavior

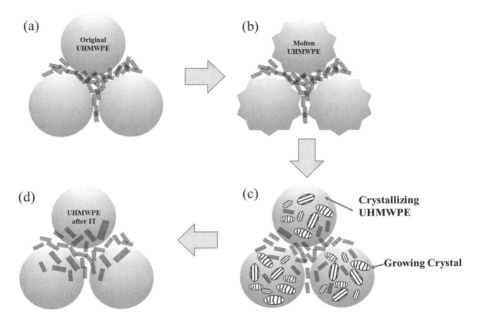

FIGURE 5.2 Schematic of microstructural development of the two-dimensional conductive network during thermal process: (a) original conductive network, (b) GNS migration, (c) GNS stabilization, (d) optimized conductive network.

Pang et al. [35] fabricated a graphene nanosheet (GNS)/UHMWPE composite with tunable I_{PTC} by the heat annealing, and Figure 5.2 represents the schematic of the microstructural development of the relevant conductive network during this thermal process. Once the polymer matrix has melted, segregated GNS were migrating into the polymer under the action of the Brownian motion (Figure 5.2b). As soon as the composites underwent subsequent crystallization, the GNS previously diffused in the polymer melt was squeezed out of the crystalline phase and was locked in the nearby amorphous phase (Figure 5.2c). Therefore, compared with the original conductive network (Figure 5.2a), a thinner conductive network of GNS was obtained (Figure 5.2d), endowing the GNS/UHMWPE composite with a higher I_{PTC}.

5.3.2 Damage Self-Sensing Behavior

Composite materials often suffer from damage caused by the applied load, impact force, or environmental aging. Therefore, it is extremely important to monitor the damage of composite materials, particularly in safety critical applications such as aerospace or automotive sectors. Some traditional monitoring techniques require embedding or attachment of sensing elements, which brings defects in the materials or needs large-scale test equipment [44,45]. Therefore, realizing non-destructive monitoring of the damage developed in composites is a key issue in many applications, and a possible way to proceed with structural health monitoring is to use the material itself as a sensor. Damage self-sensing is an in situ resistance measurement technique, which is based on the changes in electrical current (or resistance) induced by the transformation of conductive networks in CPCs. Because of the timely signal feedback and the correspondence between the conductive network and the electrical resistance, damage self-sensing monitoring has gradually become an alternative to traditional monitoring tools.

Xiang [26] prepared HDPE/MWCNT and HDPE/GNP composites with randomly dispersed structure and segregated structure, respectively, using a solution-assisted method, and their damage self-sensing behaviors were then investigated via in situ electromechanical measurements. Figure 5.3 illustrates damage self-sensing behaviors of selected HDPE/MWCNT composites, where the relationship between tensile stress and relative resistance change (RRC = $\Delta R/R_0$) versus tensile strain is presented.

One can see in Figure 5.3a–c that damage self-sensing behaviors of all selected HDPE/MWCNT composites can be divided into three main stages: (i) prior-to-yielding state, (ii) strain softening, and (iii) necking growth. For ease of understanding, the proposed deformation mechanism of conductive networks is shown in a schematic form in Figures 5.4 and 5.5. In the first stage, all the $\Delta R/R_0$ curves increase with strain elongation until a plateau is reached. These turning points, denoted as ε_1, occur at a strain of around 20% (Table 5.2), which are closely aligned with the yield points in the stress-strain curves. As is well known, the crystal structure of HDPE is almost not changed when the tensile strain is smaller than the yielding strain. Therefore, conductive networks are mainly locked by HDPE crystals, leaving only nanotubes with a small motion realm

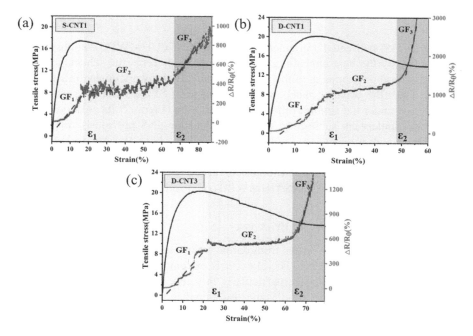

FIGURE 5.3 Tensile stress and relative resistance change $(\Delta R/R_0)$ as functions of strain in (a) S-CNT1, (b) D-CNT1, and (c) D-CNT3 composites.

Temperature-Resistivity and Damage Self-Sensing Behavior

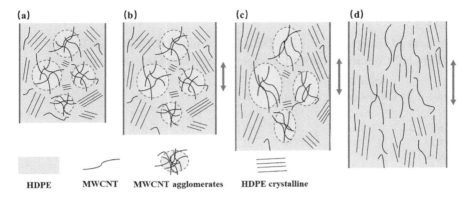

FIGURE 5.4 Schematic of the conductive network evolution in HDPE/MWCNT composites with a DCN structure: (a) before stretching, (b) prior to yielding (0 to ε_1), (c) strain softening (ε_1 to ε_2), and (d) necking growth ($> \varepsilon_2$). HDPE, high-density polyethylene; MWCNT, multi-walled carbon nanotubes.

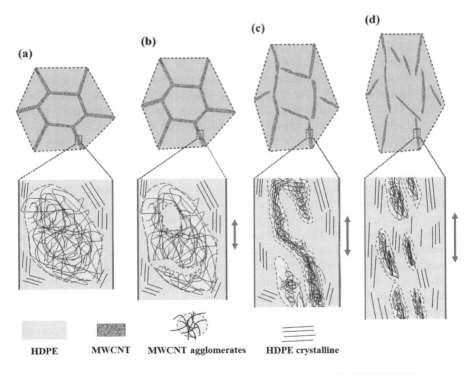

FIGURE 5.5 Schematic of the conductive network evolution in HDPE/MWCNT composites with a SCN structure: (a) before stretching, (b) prior to yielding (0 to ε_1), (c) strain softening (ε_1 to ε_2), and (d) necking growth ($> \varepsilon_2$) [26]. HDPE, high-density polyethylene; MWCNT, multi-walled carbon nanotubes.

TABLE 5.2
Summary of Gauge Factor (GF). Data Concerning HDPE/MWCNT and HDPE/GNP Composites after Damage Self-Sensing Measurements

Sample	Nanofiller Content (wt%)	$0 \sim \varepsilon_1^a$ (%)	GF_1	$0 \sim \varepsilon_2^b$ (%)	GF_2	$0 \sim \varepsilon_3^c$ (%)	GF_3
S-CNT1	1	17.8	15.8	66.8	2.2	82.5	27.3
D-CNT1	1	20.9	43.3	47.9	15.1	56.1	403.3
D-CNT3	3	21.6	21.3	62.7	3.5	72.6	89.6
S-GNP3	3	9.4	23.5	-	-	-	-
S-GNP4	4	6.2	11.5	-	-	-	-
D-GNP12	12	3.8	12.2	-	-	-	-

Note: [a-c] Representing ε_1, ε_2, and ε_3, respectively.

GF, gauge factor; CNT, carbon nanotube; GNP, graphene nanoplatelet; HDPE, high-density polyethylene; MWCNT, multi-walled carbon nanotubes.

such as increasing the tunneling distance (D_t) of neighboring nanotubes (see Figures 5.4b and 5.5b). With regard to the gauge factor in the first stage (GF_1), this value of D-CNT1 (r-CPC with 1 wt% MWCNT) was estimated to be 43.3, being about twice that of D-CNT3 (21.3). It can be attributed to a robust conductive network in the D-CNT3 system which was filled with more CNTs (3 wt% MWCNT). In addition, S-CNT1 had quite a low GF_1 value (15.8), which was even lower than that of D-CNT3 in spite of the higher content of MWCNTs. This is because the denser conductive pathways in S-CNT1, formed as MWCNTs, were mainly accumulating at the interfaces of the HDPE phases, thereby better withstanding the applied strain, which resulted in a lower GF_1.

As the strain increases from ε_1 to ε_2, the stress gradually declines and then levels off, and the relevant stage is commonly referred to as strain softening [46]. In this stage, all $\Delta R/R_0$ curves tend toward a "plateau", which means the deformation of conductive networks may reach a balance between (i) the destruction of previous conductive pathways induced by the nanotube orientation and (ii) the reconstruction of new conductive pathways resulting from other new contacts between nanotubes (Figures 5.4c and 5.5c). This is because the more obvious orientation of polymer crystals happens based on a significant increase in Hammer's factor (f_c) value, an indicative of the orientation degree of the HDPE crystals, at which the specimens are further stretched up to ε_2 (Figure 5.6a). Furthermore, a decrease of about 5% in X_{XRD} (Figure 5.6b) supports the point that the crystals are being destroyed as the deformation increases. Therefore, such a dynamically changing conductive network, as a result of further orientation of polymer crystals, basically maintains the previous conductive ability, and small variations in $\Delta R/R_0$ values (the plateau) are observed. In Figure 5.3b, D-CNT1 shows the highest $\Delta R/R_0$ value of about 1,100 at the plateau due to a less stable conductive network (Figure 5.4b). This is almost three and two times as much as those of S-CNT1 (340) and D-CNT3 (540), respectively.

At the end of the strain softening stage (denoted as ε_2), necking occurs and extends along the tensile direction with further stretching. In this "necking growth" stage,

Temperature-Resistivity and Damage Self-Sensing Behavior

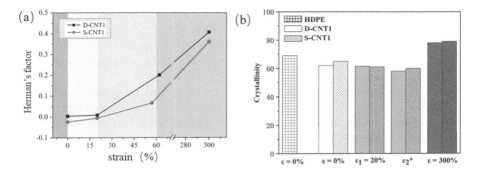

FIGURE 5.6 (a) Herman's factors (f_c) of D-CNT1 and S-CNT1 composites versus strain. (b) Crystallinity of HDPE and HDPE-based composites with different tensile strains. HDPE, high-density polyethylene.

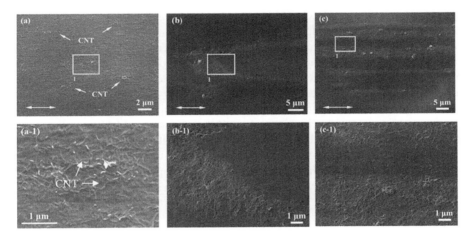

FIGURE 5.7 SEM micrographs of HDPE/MWCNT composites with different conductive networks after stretching to a strain of 300%: (a) D-CNT1, (b, c) S-CNT1. The double-headed arrows represent the tensile direction. (a-1), (b-1), and (c-1) correspond to the enlarged areas of (a), (b), and (c), respectively. HDPE, high-density polyethylene; MWCNT, multi-walled carbon nanotubes; SEM, scanning electron microscope.

the f_c (Figure 5.6a) and X_{XRD} (Figure 5.6b) values of both D-CNT1 and S-CNT1 clearly increase. This indicates that HDPE crystals are further orientated and strain-induced recrystallization occurs. In addition, the previous balance in destruction and rebuilding of conductive networks is broken with a further orientation of nanotubes, among which the destruction effect is predominant. Therefore, all $\Delta R/R_0$ curves start to increase again. One can see in Table 5.2 that D-CNT1 and S-CNT1 exhibit the highest (403.3) and lowest (27.3) GF_3 values, respectively, whereas D-CNT3 shows a moderate GF_3 value of 89.6. For the D-CNT1 sample, numerous nanotubes are disentangled and rearranged along the tensile direction (Figure 5.7a). Some neighboring nanotubes could be further separated without the ability to effectively form conductive pathways. For the S-CNT1 system, the original net-like conductive networks have been largely deformed and disrupted (Figure 5.7b), but the nanotubes are still tightly entangled at the interfaces of the HDPE phases (Figure 5.7c). Such a conductive network structure enables better retention of its electrical conductivity and results in a relatively low GF_3.

In addition to monitoring the transformation of conductive networks and crystal phases in thermoplastic composites, the damage self-sensing technique is also adapted to structural thermosetting materials such as fiber-reinforced composites. For instance, Wang et al. [47] investigated the strain and damage self-sensing capabilities of basalt fiber-reinforced polymer (BFRP) laminates fabricated with carbon nanofibers (CNFs)/epoxy composites. In particular, different contents of CNFs were dispersed into the epoxy matrix and infused into a basalt fiber fabric to form the conductive networks in BFRP laminates. According to the results, BFRP laminates containing 1.0 and 1.5 wt% CNF contents had good damage self-sensing ability to monitor the damage evolution under monotonic tensile loading until failure. Their fractional change in resistivity (FCR)-strain curves, representing the relationship between the FCR and the applied strains, could be classified into three stages corresponding to different damage modes, i.e., matrix microcracks, transverse cracks, interfacial debonding, longitudinal splitting, delamination, and basalt fiber breakage. Meng et al. [48] reported a facile approach to detect structure damages and to accurately identify their locations by using electrically conductive epoxy/graphene nanocomposite films. To this end, each edge of the composite film was equally subdivided into four points. Each point was connected to one end of the wire using silver paste, whereas the other end of the wire was coupled with the testing instrument. In the process of testing, the resistance change at each of the two opposite points was monitored, and the result that resistance change corresponding to locations of test points could point accurately locations of damages such as holes or cracks.

5.4 CONCLUSIONS

Conductive networks in CPCs are formed by directly contacted conductive fillers or by means of the tunneling current between nearby conductive fillers where electrons can flow through the insulating barriers via quantum mechanical tunneling. In most cases, CPCs can be categorized into the following three kinds with different conductive network structures: r-CPCs, s-CPCs, and d-CPCs. The r-CPCs is quite a conventional distribution where conductive fillers are randomly dispersed, and some ordinary preparing routes, e.g., melt blending and solution mixing, allow one to achieve this structure. As for s-CPCs and d-CPCs materials, their conductive fillers are both specially distributed. In the former system, fillers are primarily located at the interfaces between the matrix phases. And in the latter one, conductive fillers are selectively placed in one phase of co-continuous polymer blends. Due to these special distributions, lower contents of fillers are required to induce the insulator/conductor transition.

Some external stimuli such as temperature fluctuation and applied stress can induce the transformation of conductive networks, consequently resulting in the varying resistance of CPCs. With regard to the temperature-resistivity behavior, the PTC effect is thought to be related to the volume expansion of matrix, induced by the volume expansion of solid matrix or by the melting behavior. In contrast, the NTC effect is associated with the reconstruction of effective conductive networks. To maintain accurate responses toward the temperature stimuli, the temperature sensor is required to have a low I_{NTC} value. To this end, constructing segregated conductive networks and choosing a high melt viscosity polymer matrix are two effective approaches. In addition, high I_{PTC} is also desirable for temperature sensors because of the need for the obvious response to the variation of temperature.

In case of insulating matrices, adding appropriate amounts of conductive fillers to form an electrically conductive network has been found to be a promising approach to realize their damage self-sensing. Damage self-sensing refers to an in situ resistance measurement technique, which is based on the changes in electrical current (or resistance) induced by the transformations of conductive networks owing to the deformation or damage of CPCs. For the HDPE/MWCNT composites, three typical stages corresponding to the prior-to-yielding region, strain softening, and necking growth were mirrored by the changes in $\Delta R/R_0$ during tensile deformation. Moreover, the HDPE/MWCNT system with randomly dispersed structure was more sensitive to tensile deformation (high GF values) due to the weaker conductive networks with insufficient connections between nanotubes/agglomerates. Besides, this damage self-sensing technique is usually adapted to the structural thermosetting materials such as fiber-reinforced composites.

REFERENCES

1. Amjadi M, Kyung KU, Park I, Sitti M. Stretchable, skin-mountable, and wearable strain sensors and their potential applications: A review. *Adv Funct Mater.* 2016;26(11):1678–98.
2. Yim YJ, Park SJ. Electromagnetic interference shielding effectiveness of high-density polyethylene composites reinforced with multi-walled carbon nanotubes. *J Ind Eng Chem.* 2015;21:155–7.
3. Shi J, Li X, Cheng H, Liu Z, Zhao L, Yang T, Dai Z, Cheng Z, Shi E, Yang L, Zhang Z, Cao A, Zhu H, Fang Y. Graphene reinforced carbon nanotube networks for wearable strain sensors. *Adv Funct Mater.* 2016;26(13):2078–84.
4. Lan W, Chen Y, Yang Z, Han W, Zhou J, Zhang Y, Wang J, Tang G, Wei Y, Dou W, Su Q, Xie E. Ultraflexible transparent film heater made of Ag nanowire/PVA composite for rapid-response thermotherapy pads. *ACS Appl Mater Interfaces.* 2017;9(7):6644–51.
5. Chou TW, Gao L, Thostenson ET, Zhang Z, Byun JH. An assessment of the science and technology of carbon nanotube-based fibers and composites. *Compos Sci Technol.* 2010;70(1):1–9.
6. Zhao S, Li G, Liu H, Dai K, Zheng G, Yan X, Liu C, Chen J, Shen C, Guo Z. Positive Temperature Coefficient (PTC) Evolution of segregated structural conductive polypropylene nanocomposites with visually traceable carbon black conductive network. *Adv Mater Interfaces.* 2017;4(17):1700265
7. Na WJ, Byun JH, Lee MG, Yu WR. In-situ damage sensing of woven composites using carbon nanotube conductive networks. *Compos Part A Appl Sci Manuf.* 2015;77:229–36.
8. Wang Y, Chang R, Chen G. Strain and damage self-sensing properties of carbon nanofibers/carbon fiber-reinforced polymer laminates. *Adv Mech Eng.* 2017;9(2):1–11.
9. Duan L, Fu S, Deng H, Zhang Q, Wang K, Chen F, Fu Q. The resistivity-strain behavior of conductive polymer composites: Stability and sensitivity. *J Mater Chem A.* 2014;2(40):17085–98.

10. Zhang H, Liu Y, Kuwata M, Bilotti E, Peijs T. Improved fracture toughness and integrated damage sensing capability by spray coated CNTs on carbon fibre prepreg. *Compos Part A Appl Sci Manuf.* 2015;70:102–10.
11. Bai YL, Dai JG, Teng JG. Buckling of steel reinforcing bars in FRP-confined RC columns: An experimental study. *Constr Build Mater.* 2017;140: 403–15.
12. Bauhofer W, Kovacs JZ. A review and analysis of electrical percolation in carbon nanotube polymer composites. *Compos Sci Technol.* 2009; 69(10):1486–98.
13. Xu S, Rezvanian O, Peters K, Zikry MA. The viability and limitations of percolation theory in modeling the electrical behavior of carbon nanotube-polymer composites. *Nanotechnology.* 2013;24(15):155706.
14. Zhang R, Baxendale M, Peijs T. Universal resistivity-strain dependence of carbon nanotube/polymer composites. *Phys Rev B.* 2007;76(16):195433.
15. Nanni F, Mayoral BL, Madau F, Montesperelli G, McNally T. Effect of MWCNT alignment on mechanical and self-monitoring properties of extruded PET-MWCNT nanocomposites. *Compos Sci Technol.* 2012;72(10):1140–6.
16. Xiang D, Wang L, Tang Y, Hill CJ, Chen B, Harkin-Jones E. Reinforcement effect and synergy of carbon nanofillers with different dimensions in high density polyethylene based nanocomposites. *Int J Mater Res.* 2017;108(4):322–34.
17. Liu H, Gao J, Huang W, Dai K, Zheng G, Liu C, Shen C, Yan X, Guo J, Guo Z. Electrically conductive strain sensing polyurethane nanocomposites with synergistic carbon nanotubes and graphene bifillers. *Nanoscale.* 2016;8(26): 12977–89.
18. Sichel EK, Sheng P, Gittleman JI, Bozowski S. Observation of fluctuation modulation of tunnel junctions by applied ac stress in carbon polyvinylchloride composites. *Phys Rev B.* 1981;24(10):214304.
19. Sengupta R, Bhattacharya M, Bandyopadhyay S, Bhowmick AK. A review on the mechanical and electrical properties of graphite and modified graphite reinforced polymer composites. *Progr Polym Sci.* 2011;36(5):638–70.
20. Ahir SV, Huang YY, Terentjev EM. Polymers with aligned carbon nanotubes: Active composite materials. *Polym.* 2008;49(18):3841–54.
21. Al-Saleh MH, Sundararaj U. A review of vapor grown carbon nanofiber/polymer conductive composites. *Carbon.* 2009;47(1):2–22.
22. Spitalsky Z, Tasis D, Papagelis K, Galiotis C. Carbon nanotube-polymer composites: Chemistry, processing, mechanical and electrical properties. *Progr Polym Sci.* 2010;35(3):357–401.
23. Villmow T, Kretzschmar B, Poetschke P. Influence of screw configuration, residence time, and specific mechanical energy in twin-screw extrusion of polycaprolactone/multi-walled carbon nanotube composites. *Compos Sci Technol.* 2010;70(14):2045–55.
24. Sahoo NG, Rana S, Cho JW, Li L, Chan SH. Polymer nanocomposites based on functionalized carbon nanotubes. *Progr Polym Sci.* 2010;35(7):837–67.
25. Ajayan PM, Stephan O, Colliex C, Trauth D. Aligned carbon nanotube arrays formed by cutting a polymer resin--nanotube composite. *Science (New York, NY).* 1994;265(5176):1212–4.
26. Xiang D, Wang L, Tang Y, Harkin-Jones E, Zhao C, Wang P, Li Y. Damage self-sensing behavior of carbon nanofiller reinforced polymer composites with different conductive network structures. *Polymer.* 2018;158:308–19.
27. Ghislandi M, Tkalya E, Marinho B, Koning CE, de With G. Electrical conductivities of carbon powder nanofillers and their latex-based polymer composites. *Compos Part A Appl Sci Manuf.* 2013;53:145–51.
28. Chen J, Shi YY, Yang JH, Zhang N, Huang T, Chen C, Wang Y, Zhou Z.-W. A simple strategy to achieve very low percolation threshold via the selective distribution of carbon nanotubes at the interface of polymer blends. *J Mater Chem.* 2012;22(42): 22398–404.

29. Gubbles F, Blacher S, Vanlathem E. Design of electrical conductive composites: Key role of the morphology on the electrical properties of carbon black filled polymer blends. Macromolecules. 1995;28(5):1559–1566.
30. Zhang S, Deng H, Zhang Q, Fu Q. Formation of conductive networks with both segregated and double-percolated characteristic in conductive polymer composites with balanced properties. *ACS Appl Mater Interfaces*. 2014;6(9):6835–44.
31. Pang H, Yan DX, Bao Y, Chen JB, Chen C, Li ZM. Super-tough conducting carbon nanotube/ultrahigh-molecular-weight polyethylene composites with segregated and double-percolated structure. *J Mater Chem*. 2012;22(44):23568–75.
32. Xu Z, Zhang Y, Wang Z, Sun N, Li H. Enhancement of electrical conductivity by changing phase morphology for composites consisting of polylactide and poly(epsilon-caprolactone) filled with acid-oxidized multiwalled carbon nanotubes. *ACS Appl Mater Interfaces*. 2011;3(12):4858–64.
33. Goeldel A, Kasaliwal G, Poetschke P. Selective localization and migration of multiwalled carbon nanotubes in blends of polycarbonate and poly(styrene-acrylonitrile). *Macromol Rapid Commun*. 2009;30(6):423–9.
34. Zhang L, Wan C, Zhang Y. Morphology and electrical properties of polyamide 6/polypropylene/multi-walled carbon nanotubes composites. *Compos Sci Technol*. 2009;69(13):2212–7.
35. Pang H, Zhang YC, Chen T, Zeng BQ, Li ZM. Tunable positive temperature coefficient of resistivity in an electrically conducting polymer/graphene composite. *Appl Phys Lett*. 2010;96(25):251907.
36. Xiang D, Wang L, Tang Y, Zhao C, Harkin-Jones E, Li Y. Effect of phase transitions on the electrical properties of polymer/carbon nanotube and polymer/graphene nanoplatelet composites with different conductive network structures. *Polym Int*. 2018;67(2):227–35.
37. Deng H, Skipa T, Zhang R, Lellinger D, Bilotti E, Alig I, Peijs T. Effect of melting and crystallization on the conductive network in conductive polymer composites. *Polymer*. 2009;50(15):3747–54.
38. Deng H, Lin L, Ji M, Zhang S, Yang M, Fu Q. Progress on the morphological control of conductive network in conductive polymer composites and the use as electroactive multifunctional materials. *Progr Polym Sci*. 2014;39(4):627–55.
39. Dai K, Li ZM, Xu XB. Electrically conductive in situ microfibrillar composite with a selective carbon black distribution: An unusual resistivity-temperature behavior upon cooling. *Polymer*. 2008;49(4):1037–48.
40. Dai K, Zhang YC, Tang JH, Ji X, Li ZM. Anomalous attenuation and structural origin of positive temperature coefficient (PTC) effect in a carbon black (CB)/poly(ethylene terephthalate) (PET)/polyethylene (PE) electrically conductive microfibrillar polymer composite with a preferential CB distribution. *J Appl Polym Sci*. 2012;125:E561–70.
41. Gao JF, Li ZM, Peng S, Yan DX. Temperature-resistivity behaviour of CNTs/UHMWPE composites with a two-dimensional conductive network. *Polym-Plast Technol Eng*. 2009;48(4):478–81.
42. Lu C, Cao QQ, Huang XH, Hu XN, He YX, Liu CY, Zhang YQ. Influence of morphology on positive temperature coefficient effect for conductive polymer composites with carbon black dispersed at interface. *Polym Eng Sci*. 2013;53(12):2640–9.
43. Zhang C, Ma CA, Wang P, Sumita M. Temperature dependence of electrical resistivity for carbon black filled ultra-high molecular weight polyethylene composites prepared by hot compaction. *Carbon*. 2005;43(12):2544–53.
44. Yeager M, Todd M, Gregory W, Key C. Assessment of embedded fiber Bragg gratings for structural health monitoring of composites. *Struct Health Monit Int J*. 2017;16(3):262–75.
45. Janapati V, Kopsaftopoulos F, Li F, Lee SJ, Chang FK. Damage detection sensitivity characterization of acousto-ultrasound-based structural health monitoring techniques. *Struct Health Monit Int J*. 2016;15(2):143–61.

46. Chen K, Schweizer KS. Theory of yielding, strain softening, and steady plastic flow in polymer glasses under constant strain rate deformation. *Macromolecules.* 2011;44(10):3988–4000.
47. Wang Y, Wang Y, Wan B, Han B, Cai G, Chang R. Strain and damage self-sensing of basalt fiber reinforced polymer laminates fabricated with carbon nanofibers/epoxy composites under tension. *Compos Part A-Appl Sci Manuf.* 2018;113:40–52.
48. Meng Q, Araby S, Oh JA, Chand A, Zhang X, Kenelak V, Ma J, Liu T, Ma J. Accurate self-damage detection by electrically conductive epoxy/graphene nanocomposite film. *J Appl Polym Sci.* 2021;138(20):50452.

6 Flexible Strain Sensors Based on Elastic Fibers of Conductive Polymer Composites

6.1 INTRODUCTION

With continuous growing living standard and economic development, humans pay more attention to physical health. Physical exercise has become an important way to keep fit in our daily life. Smart watches and wristbands acquire physiological data during exercise, but they do not enable the real-time monitoring of various parts of the body. Moreover, chronic diseases have gradually become more common in the international community due to environmental deterioration and an aging population. The human health care system is thus facing severe challenges. Large medical equipment and cumbersome medical practices not only increase the cost of health monitoring, but they also bring unnecessary anxiety and pain to patients. However, conventional medical monitoring based on metals or semiconductors cannot meet the requirements due to the rigidity and brittleness of these materials. Therefore, with the development of material science and continuous economic development, artificial intelligence technologies have been innovating rapidly. In particular, flexible strain sensors have undergone tremendous development and have been applied in robotics, medical diagnostics, human motion detection, and health care over the last decade [1–4]. Flexible fiber strain sensors have become a hot topic in academia because fiber sensors are readily incorporated into clothing or wearable equipment. Additionally, flexible strain sensors with high sensitivity and workable strain range show great potential for use in various fields.

To date, various conductive functional materials have been used to fabricate strain sensors including carbon black (CB) [5,6], carbon nanotubes (CNTs) [7,8], graphene [9,10], and metal nanowires [11,12]. Electrically conductive functional fillers incorporated into a polymer matrix and designing ingenious structures are effective strategies for improving the performance of flexible strain sensors. The performance of strain sensors must be excellent and needs to be measured. The parameters measured include workable range, sensitivity, stability, response time, hysteresis, and linear relationship between strain and change of relative resistance. Due to their easy fabrication, facile integration, and excellent performance, fibrous sensors have been designed and fabricated to detect various stimuli for many different applications. In particular, fibrous sensors can be easily woven into different textures or even knitted into fabric structures. Fabric is defined as hierarchically structured fibrous materials that assemble fibers at a two-dimensional level. Flexible fabric is the ideal platform to fabricate flexible electronic devices or wearable systems. Generally, layer-by-layer (LBL) coating [13], ultrasonic-assisted dip-coating [14], chemical deposition coating [15], and spinning [16] approaches are the main methods used to fabricate flexible fiber strain sensors. Flexible strain sensors with high workable ranges and sensitivities have potential applications in personal health care, body motion detection, and man–machine interactions.

This chapter focuses on elastic fibers of conductive polymer composite (CPC) strain sensors with good weaving capacity and enhanced stability, which is one of the most explored areas in textile sensor research. First, we review different approaches of making elastic fiber strain sensors, which include LBL coating, chemical deposition coating, melt extrusion, and spinning. Second, we discuss the sensing performance of flexible strain sensors with different structures. In this section, the working mechanisms of conductive networks with different microstructures and the sensing performances are analyzed and compared. Finally, the potential applications of flexible strain sensors and current challenges are reviewed. This review aims to present an overview of all research of elastic fibers of CPC strain sensing networks and their applications in various fields including smart wearable devices and integrated multifunctional human–machine interactive systems.

6.2 METHODS FOR FABRICATING FLEXIBLE FIBER STRAIN SENSORS

Coating of functional layers onto the surface of fibers is a widely used facile technique. This method not only has low cost and is easy to operate, but it also exhibits high efficiency. Surprisingly, coating functional materials onto the surface of various hierarchical levels of textile structure such as fiber, yarn, or fabric can be completed successfully. In addition, flexible fiber strain sensors fabricated by dip-coating, chemical deposition coating, melt extrusion, and spinning have also been reported.

6.2.1 LBL Coating and Ultrasonic-Assisted Dip-Coating

LBL coating is a low-cost, simple, and efficient approach for manufacturing flexible fiber strain sensors. In recently reported work, Chen et al. [13] prepared flexible, sensitive, and wearable strain sensors via LBL coating of thermoplastic polyurethane (TPU)/multi-walled carbon nanotube (MWCNT) composites onto the surface of spandex yarn (see Figure 6.1a). This procedure resulted in an ultrathin layer of conductive composite material on the surface of the spandex yarn. Ingeniously, this homogenous coating strategy easily avoid phase separation caused by different substrates as spandex is also a TPU. Additionally, commercially available spandex yarn can be readily incorporated into clothing or wearable equipment for further applications.

The LBL method is widely used in the preparation of flexible fiber strain sensors because of its many advantages. However, the mechanism of LBL coating is different. For example, Wu et al. [1] prepared flexible, sensitive, and washable strain sensors based on polyurethane (PU) yarn using an LBL coating technique by alternately dipping PU yarn into a suspension containing negatively charged carbon blank (CB)/cellulose

Flexible Fiber Strain Sensors

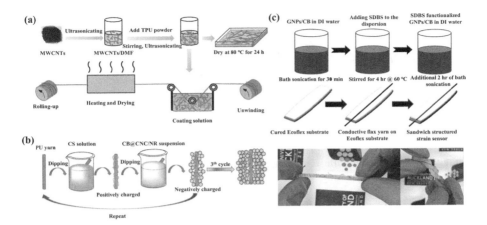

FIGURE 6.1 Schematic illustration of the fabrication of a flexible strain sensor by layer-by-layer (LBL) coating or ultrasonically assisted dip-coating approaches. (a) Fabrication of flexible yarn strain sensor by LBL coating, (b) schematic process for the fabrication of CPC@PU yarn by LBL assembly, and (c) preparation of conductive yarn by ultrasonically assisted dip-coating.

nanocrystals/natural rubber suspension and positively charged chitosan solution as shown in Figure 6.1b. Specifically, a functional material was coated on the surface of PU yarn to form an ultrathin CPC layer via the interaction of positive and negative charges. The resultant flexible yarn strain sensor exhibited excellent sensitivity with a gauge factor (GF) of 39 and detection limit of 0.1% strain as well as good reproducibility over 10,000 cycles. The developed flexible strain sensor exhibited good flexibility and stretchability, and it could easily be incorporated into textile structures through weaving, knitting, and braiding for wearable sensing applications.

In addition, ultrasonic-assisted LBL processing can effectively solve problems of agglomeration of nanofillers in matrices. Souri and Bhattacharyya [17] fabricated conductive yarn via dip-coating with graphene nanoplatelets (GNPs) and CB particles in an ultrasonication bath (see Figure 6.1c). Under ultrasonication, GNP and CB were uniformly dispersed on the surface of yarn. At the same time, the hierarchical-conductive network was successfully established. In summary, the ultrasonically assisted LBL method is effective in improving the decentralization of conductive nanofillers for the flexible application in the fabrication of yarn strain sensors. The developed yarn strain sensor exhibits good flexibility and stretchability, which can be easily incorporated into textile structures through weaving, knitting, and braiding for wearable sensing applications.

6.2.2 CHEMICAL DEPOSITION COATING

The LBL method and ultrasonically assisted dip-coating to prepare flexible strain sensors have many important advantages, but it cannot meet all of the requirements of the preparation of flexible strain sensors. In particular, there are still some challenges such as the dispersion of nanoparticles in the polymer matrix and weak

116 Carbon-Based Conductive Polymer Composites

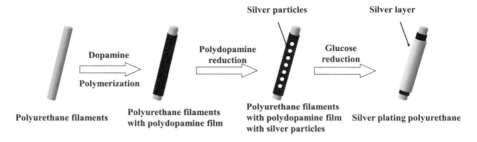

FIGURE 6.2 Schematic illustration of conductive polyurethane filaments by in situ reduction and electroless Ag plating.

interfacial interactions between the filler and the matrix. However, chemical deposition coating is an effective technique to achieve homogenous integration between inorganic components and the polymer matrix. Liu et al. [18] developed an in situ polymerization method to prepare an adherent polydopamine (PDA) film on PU filaments for deposition of Ag particles as shown in Figure 6.2. The Ag nanoparticles were able to deposit well because of the catechol groups of PDA. The developed flexible yarn strain sensor with high elasticity and linearity was successfully applied to wearable strain sensing devices. Therefore, chemical deposition coating is an effective approach to deposit functional conductive components on flexible yarn. In addition, the chemical bonding can improve the adhesion ability between the deposition layer and yarn surface.

6.2.3 MELT EXTRUSION

Improvement in the strength and durability of flexible strain sensors are scientifically and technologically necessary to meet the application of wearable devices. To prepare flexible strain sensors with stable conductive networks, the polymer matrix and functional materials were integrated by melt extrusion. Lin et al. [19] reported the construction of a polypropylene (PP)/CNT composite layer that was attached to a polyester fiber via a melt extrusion method. The polyester fiber provided robust mechanical properties for the conductive composite fiber material. Recently, Liao et al. [20] proposed a cluster-type microstructure strategy for the fabrication of yarn strain sensors using a nozzle jet printing method (see Figure 6.3). This work

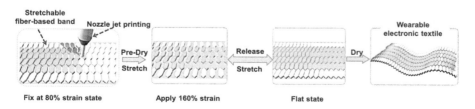

FIGURE 6.3 Fabrication of strain sensors with cluster-type microstructures.

Flexible Fiber Strain Sensors

demonstrates that the intrinsic elasticity of textiles can be used to realize unique microstructure designs with nozzle jet printing of the conductive layer. The melt extrusion approach is also favorable because of its excellent sample formability and structure designability.

6.2.4 Spinning

Spinning techniques have been widely used to fabricate fiber. Generally, spinning processes to fabricate fiber sensors include the following steps. First, the spinning solution is prepared. Second, the spinning solution is extruded from the spinneret. Third, primary fibers are immersed in the solidifying solution. Fourth, a post-treatment step is performed [21,22]. Therefore, spinning technique is an easy method to fabricate flexible fiber strain senors. For instance, Seyedin et al. [23] developed an electrically conductive and highly stretchable PU/PEDOT:PSS yarn by spinning. The PU/PEDOT:PSS yarn exhibited robust mechanical properties to meet the requirements of the knitting technique. Recently, He et al. [24] reported a novel highly sensitive strain sensor based on MWCNTs and TPU through a wet spinning process in Figure 6.4. The spinning approach not only can be used to effectively prepare conductive elastomeric fiber strain sensors, but the structure of the fiber can also be easily designed by facile alteration of loop configuration and stitch insertion.

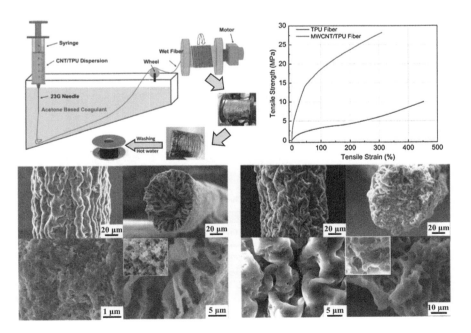

FIGURE 6.4 Schematic illustration of the wet spinning process for a MWCNT/TPU fiber strain sensor. MWCNT, multi-walled carbon nanotube; TPU, thermoplastic polyurethane.

6.3 RELATIONSHIP BETWEEN STRUCTURE AND PERFORMANCE OF FLEXIBLE FIBER STRAIN SENSOR

In this section, the three different classes of fibrous materials used to prepare strain sensors are summarized. These classes are sheath-core spun yarn, helical yarn, and fabric. In addition, the effects of the different classes of fibrous structures on sensing performance are also discussed.

6.3.1 Sheath-Core Spun Yarn

Multiple mechanical deformation sensors have posed an urgent need for industrial application. Chen et al. [25] developed a facile and low-cost approach to fabricate spandex strain sensors using TPU/MWCNTs composites as shown in Figure 6.5a. The sheath-core spun spandex consists of multiple monofilaments tightly bonded, which exhibit excellent sensing performance after coating a conductive composite to obtain electrical conductivity. In another work, commercial composite yarn consisting of central elastic rubber latex thread and winding PU fibers were used as scaffold. The stretched yarn was deposited with P(VDF-TrFE) nanofibers followed by the deposition of Ag as a conductive layer [26]. Wang et al. [27] designed a wrapping and coating device to achieve the fabrication of a cotton/PU core-spun yarn. During the winding process of cotton fibers on the PU filament surface, conductive single-wall CNTs were incorporated into the core-spun yarn through a coating treatment. The self-designed equipment was simple and achieved uniform covering of the twisted fibers and resulted in the scalable production of sheath-core yarn [28]. Considering

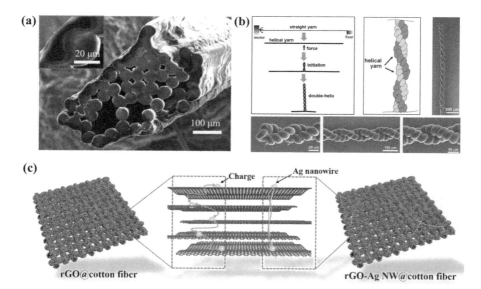

FIGURE 6.5 Illustration of the different dimensions of fiber strain sensors. (a) Yarn strain sensor, (b) fabrication and characterization of double-helix carbon nanotube yarn, and (c) flexible fabric strain sensor.

Flexible Fiber Strain Sensors 119

the brittleness of carbon nanofiber yarn, the generated subtle cracks can increase the sensitivity during the stretching process. Yan et al. [29] reported the fabrication of sheath-core helical yarn through carbonization of core cotton yarn, and electrospun polyacrylonitrile nanofiber was used as a wrapping sheath. The yarn was effective at monitoring subtle strains as low as 0.1% with good sensitivity. Recently, natural silk fiber was also functionalized by tailor-made CNT paint to fabricate strain sensors, and this device was used to detect the physical stimuli of human body [30].

6.3.2 Helical Yarn

Because of its helically twisted filament, the compression spring structure is an effective approach to obtain good elasticity. Different from the abovementioned sheath-core structure, helical structures can be fabricated using a single yarn. For example, Zhao et al. [31] developed a prototype CNT yarn strain sensor with excellent repeatability and stability for in situ structural health detection. The yarn was directly spun from as-grown CNT arrays, and the twisting process resulted in an electrically conductive pathway in the longitudinal direction. This device is a promising strain sensor as the electrical resistance increases linearly with tensile strain. Shang et al. [32] fabricated yarn-derived spring-like CNT rope consisting of uniformly arranged loops. The spring-like rope was obtained by overtwisting the randomly oriented CNT film using a modified spinning technique. Furthermore, Shang et al. [33] prepared yarn-derived two-level hierarchical composite structures consisting of twisted double-helical yarns as shown in Figure 6.5b. The yarn end was adaptive to recoverable drag, resulting in a large linear change of tensile strain against electrical resistance. The extensively twisted effects of entanglement enabled the yarn to function as a stretchable strain sensor [34].

6.3.3 Fabric

With increasing social development, fabric has become indispensable in people's daily life. Cotton fiber can be obtained from many different sources, is inexpensive, exhibits excellent performance, and is environmentally friendly. For these reasons, cotton is one of the most popular natural fibers in the clothing and textile field. Weaving and knitting can be used to make cotton fabrics with yarns and fibers in the form of a hierarchical structure with good tensile properties. There were some studies focused on coating various types of conductive materials on the surface of cotton fabrics including reduced graphene oxide (rGO) [35] and MWCNTs [36], which only used cotton fabrics as flexible substrates. Souri and Bhattacharyya [37] mixed GNPs and CB particles to cover the surface of the cotton fabrics. Through a pre-breaking mechanism, a part of the fabrics was first broken, and resistance changes became more sensitive to fiber breaking under high strain, resulting in an ultra-high GF of 102350.91. These composite materials all possessed excellent sensitivity and stability in motion detection of human fingers. Cao et al. [38] found that the conductivity of the cotton fabric was not satisfactory when covering it with pure graphene due to the poor performance of resistance between the layers of graphene, so the authors coated the graphene with AgNW to improve conductivity as shown in Figure 6.5c. In summary, the strain range and ductility of the sensor can be significantly improved by using different structures of fabric as flexible substrates, while the design of conductive network structure affects the sensitivities of the sensors.

The use of an appropriate base substrate plays a vital role for the production of conductive textiles. Depending on the type of sensor, many researchers used 100% cotton woven (plain and twill) fabrics rather than knit fabrics due to extensibility characteristics and the greater porosity of knit fabric. Lower porosity of woven fabric is more suitable for control of the ink during inkjet printing. Moreover, woven fabrics have greater dimensional stability than knit fabrics. Woven fabrics are less easily deformed during the laundering process depending on the fiber content, weaving structure, and washing cycle.

6.4 APPLICATIONS OF FLEXIBLE FIBER STRAIN SENSORS

Flexible strain sensors based on CPCs exhibit high working ranges and sensitivities. For these reasons, they have become an ideal element to use in wearable device and other electronic systems. There are three applications for flexible fiber strain sensors based on CPCs. These three applications are personal health care, body motion detection, and man–machine interactions. The flexible fiber strain sensors are more advantageous in these fields compared with sensors based on traditional metals and semiconductors due to their high flexibility and deformation ability. Additionally, flexible fibers can be readily incorporated into clothing or wearable equipment for various applications.

6.4.1 Personal Health Care

In recent years, the harm of chronic diseases to human health has generated widespread concern. Therefore, health care is also becoming a hot topic. Pulse, blood pressure, and the long-term monitoring of respiratory rate are very helpful for doctors to accurately diagnose chronic diseases. Obviously, traditional monitoring equipment is too bulky and heavy to be suitable for long-term wearing. However, the deformation generated by flexible CPC fibers leads to a change in resistance. Furthermore, electrical signals of the flexible strain sensors can distinguish the various actions stimulated. For instance, flexible strain sensors have been mounted in the throat, wrist, and abdomen to ascertain different physiological information [39]. Flexible fiber strain sensors have demonstrated high sensitivity and excellent stability, and they exhibit an accurate change of peak resistance under deformation [25] as shown in Figure 6.6a. Flexible and wearable technology will play an important role in the field of health monitoring in the future due to advantages of easy implementation, early diagnosis, and long-term monitoring.

6.4.2 Body Motion Detection

Physical exercise is a good thing for people's health. However, inappropriate physical exercise causes injury or even more serious consequences. Currently, the flexible fiber strain sensors can greatly reduce injury during physical exercise by monitoring and adjusting exercise real time. Furthermore, flexible fiber sensors assist training, which

Flexible Fiber Strain Sensors

is particularly important for professional athletes. The training schemes of athletes can be adjusted based on the signals of flexible strain sensors for improved performance. Additionally, body motion has also been monitored by flexible fiber strain sensors including facial expressions and joint motions. For instance, a flexible fiber strain sensor was mounted on the forehead and cheek to capture signals of expression, including frowning, blinking, smiling, and crying. The sensor was first attached on the person's throat. When the person reads different English words, the strain sensor showed a distinct signal pattern for each word attributed to the specific muscle movement and subtle pressure changes needed to pronounce each word. As shown in Figure 6.6b, the different feedback signals generated by walking, jogging, jumping, and squatting were distinguishable [40]. This work demonstrates good sensing performance of the flexible fiber strain sensors in monitoring delicate movements of human muscles. In summary, these results suggest that these sensors are quite useful for monitoring the body healthy condition and evaluating athletes' sport performance.

6.4.3 Human–Machine Interactions

As a potential application, the flexible fiber strain sensors are prospective to build a smart human–machine interface system to control machines. The flexible sensors

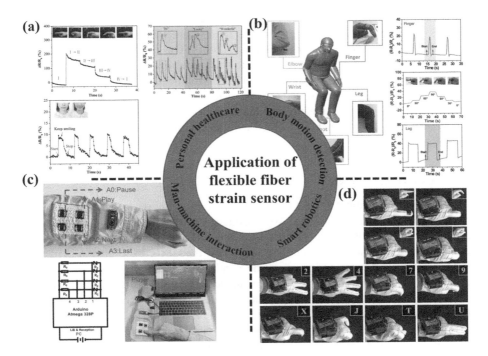

FIGURE 6.6 Applications of flexible fiber strain sensors based on conductive polymer composites in the fields of personal health care, body motion detection, man–machine interactions, and smart robotics monitoring. (a) Personal health care, (b) body motion detection, (c) man–machine interactions, and (d) smart robotics.

could capture the deformation of human activities, location, contact force, and processes into sensing signals, which could be treated by software and then sent to machines. Signals acquired from the sensor can also be used to control smart robots. As shown in Figure 6.6c, flexible strain sensors based on CPCs could be used to develop intelligent gloves, input signals by finger bending to the controller for remote control of smart robots, and allow them to perceive the environment and interact with environmental goals. These devices could reduce the deviation through real-time feedback to perform surgical operations or some fine and dangerous works that humans could not complete [41]. Due to the high sensitivity, low fabrication cost, and simple operation, the smart glove system developed with flexible strain sensors based on CPC has more advantages over traditional systems using sensors based on metals and semiconductors.

6.4.4 INTELLIGENT ROBOTICS

Strain sensors as sensing elements were incorporated into machines to develop smart robotics. As shown in Figure 6.6d, a sensory array was designed to differentiate the pressing positions and the magnitude of force on gloves [42]. When a finger touched the sensor, the fast response on a cylindrical object was formed to show the pressing area. Moreover, multiple locations could be monitored simultaneously by the sensing array. The different gestures of a finger could also be detected. These sorts of devices are promising as sensing skins for robots. The strain sensor could also be used to analyze the gait in the finger of a smart robot. The exerted force of the robot's finger during working could be recorded to track physical movements of the robot. The sensing information provided a model to control the activity of robot. Similarly, the fingers could also be controlled to grab various objects. These flexible strain sensors will be able to control the physical movement of robots with the improvement of technology integration capability, which will then helps humans with a wide variety of dangerous tasks.

6.5 CONCLUSIONS

In this chapter, a critical review has been gathered on the recent advances of elastic fibers of CPC strain sensors. We introduced in detail the various fabrication approaches of flexible strain sensors including LBL coating, chemical deposition coating, melt extrusion, and spinning. Generally, the CPCs are prepared by depositing a conductive layer or incorporating conductive functional materials into the polymer matrix. Next, we discuss the sensing performances of flexible strain sensors with different structures. The unique sheath-core structure is beneficial to increase both elasticity and mechanical strength. In addition, the helically twisted filament is also an effective approach to obtain good elasticity. Woven fabrics possess greater dimensional stability than single fiber or yarn. Moreover, woven fabrics are less easily deformed by laundering and washing processes. Finally, the applications of flexible fiber strain sensors were summarized. Overall, material selection and structure design remain major challenges for the fabrication of flexible CPC strain sensors. The sensors combine high-performance conductive materials with polymer composite flexible substrates with good plasticity. Researchers tend to prepare these materials by simple mixing or coating, but few studies have been conducted on the interaction between them. Furthermore, the inability to simultaneously obtain high tensile properties and sensitivity is a common challenge in this research area. Achieving higher sensitivity in the strain range needed for human motion detection requires further structural optimization. In the future, fiber-shaped strain with novel structures and sensing efficiencies will be designed and further engineered to develop scalable, highly sensitive, stretchable, durable, and reliable wearable electronic devices.

REFERENCES

1. Wu XD, Han YY, Zhang XX, Lu CH. Highly sensitive, stretchable, and wash-durable strain sensor based on ultrathin conductive layer@polyurethane yarn for tiny motion monitoring. *ACS Appl Mater Inter.* 2016;8(15):9936–45.
2. Zhang BC, Wang H, Zhao Y, Li F, Ou XM, Sun BQ, et al. Large-scale assembly of highly sensitive Si-based flexible strain sensors for human motion monitoring. *Nanoscale.* 2016;8(4):2123–8.
3. Ryu S, Lee P, Chou JB, Xu RZ, Zhao R, Hart AJ, et al. Extremely elastic wearable carbon nanotube fiber strain sensor for monitoring of human motion. *ACS Nano.* 2015;9(6):5929–36.
4. Heo JS, Eom J, Kim YH, Park SK. Recent progress of textile-based wearable electronics: A comprehensive review of materials, devices, and applications. *Small.* 2018;14(3):1703034.
5. Zheng YJ, Li YL, Li ZY, Wang YL, Dai K, Zheng GQ, et al. The effect of filler dimensionality on the electromechanical performance of polydimethylsiloxane based conductive nanocomposites for flexible strain sensors. *Compos Sci Technol.* 2017;139:64–73.
6. Pan YM, Liu XH, Hao XQ, Schubert DW. Conductivity and phase morphology of carbon black-filled immiscible polymer blends under creep: An experimental and theoretical study. *Phys Chem Chem Phys.* 2016;18(47):32125–31.
7. Gao LM, Thostenson ET, Zhang Z, Chou TW. Sensing of damage mechanisms in fiber-reinforced composites under cyclic loading using carbon nanotubes. *Adv Funct Mater.* 2009;19(1):123–30.
8. Chen JW, Cui XH, Zhu YT, Jiang W, Sui KY. Design of superior conductive polymer composite with precisely controlling carbon nanotubes at the interface of a co-continuous polymer blend via a balance of pi-pi interactions and dipole-dipole interactions. *Carbon.* 2017;114:441–8.
9. Cheng Y, Wang RR, Sun J, Gao L. A stretchable and highly sensitive graphene-based fiber for sensing tensile strain, bending, and torsion. *Adv Mater.* 2015;27(45):7365–71.
10. Jang H, Park YJ, Chen X, Das T, Kim MS, Ahn JH. Graphene-based flexible and stretchable electronics. *Adv Mater.* 2016;28(22):4184–202.
11. Yao SS, Zhu Y. Wearable multifunctional sensors using printed stretchable conductors made of silver nanowires. *Nanoscale.* 2014;6(4):2345–52.
12. Amjadi M, Pichitpajongkit A, Lee S, Ryu S, Park I. Highly stretchable and sensitive strain sensor based on silver nanowire-elastomer nanocomposite. *ACS Nano.* 2014;8(5):5154–63.
13. Chen Q, Xiang D, Wang L, Tang YH, Harkin-Jones E, Zhao CX, et al. Facile fabrication and performance of robust polymer/carbon nanotube coated spandex fibers for strain sensing. *Compos Part A Appl S.* 2018;112:186–96.
14. Li YH, Zhou B, Zheng GQ, Liu XH, Li TX, Yan C, et al. Continuously prepared highly conductive and stretchable SWNT/MWNT synergistically composited electrospun thermoplastic polyurethane yarns for wearable sensing. *J Mater Chem C.* 2018;6(9):2258–69.

15. Bashir T, Skrifvars M, Persson NK. Synthesis of high performance, conductive PEDOT-coated polyester yarns by OCVD technique. *Polym Advan Technol*. 2012;23(3):611–7.
16. Seyedin S, Razal JM, Innis PC, Jeiranikhameneh A, Beirne S, Wallace GG. Knitted strain sensor textiles of highly conductive all-polymeric fibers. *ACS Appl Mater Inter*. 2015;7(38):21150–8.
17. Souri H, Bhattacharyya D. Wearable strain sensors based on electrically conductive natural fiber yarns. *Mater Design*. 2018;154:217–27.
18. Liu H, Zhu LL, He Y, Cheng BW. A novel method for fabricating elastic conductive polyurethane filaments by in-situ reduction of polydopamine and electroless silver plating. *Mater Design*. 2017;113:254–63.
19. Lin JH, Lin ZI, Pan YJ, Hsieh CT, Lee MC, Lou CW. Manufacturing techniques and property evaluations of conductive composite yarns coated with polypropylene and multi-walled carbon nanotubes. *Compos Part A Appl S*. 2016;84:354–63.
20. Liao XQ, Wang WS, Wang L, Tang K, Zheng YJ. Controllably enhancing stretchability of highly sensitive fiber-based strain sensors for intelligent monitoring. *ACS Appl Mater Inter*. 2019;11(2):2431–40.
21. Lundahl MJ, Klar V, Wang L, Ago M, Rojast OJ. Spinning of cellulose nanofibrils into filaments: A review. *Ind Eng Chem Res*. 2017;56(1):8–19.
22. Weisser P, Barbier G, Richard C, Drean JY. Characterization of the coagulation process: Wet-spinning tool development and void fraction evaluation. *Text Res J*. 2016;86(11):1210–9.
23. Seyedin S, Moradi S, Singh C, Razal JM. Continuous production of stretchable conductive multifilaments in kilometer scale enables facile knitting of wearable strain sensing textiles. *Appl Mater Today*. 2018;11:255–63.
24. He ZL, Zhou GH, Byun JH, Lee SK, Um MK, Park B, et al. Highly stretchable multi-walled carbon nanotube/thermoplastic polyurethane composite fibers for ultrasensitive, wearable strain sensors. *Nanoscale*. 2019;11(13):5884–90.
25. Chen Q, Li YT, Xiang D, Zheng YF, Zhu WQ, Zhao CX, et al. Enhanced strain sensing performance of polymer/carbon nanotube-coated spandex fibers via noncovalent interactions. *Macromol Mater Eng*. 2020;305(2):1900525.
26. Chen S, Lou Z, Chen D, Jiang K, Shen GZ. Polymer-enhanced highly stretchable conductive fiber strain sensor used for electronic data gloves. *Adv Mater Technol-Us*. 2016;1(7):1600316.
27. Wang ZF, Huang Y, Sun JF, Huang Y, Hu H, Jiang RJ, et al. Polyurethane/cotton/carbon nanotubes core-spun yarn as high reliability stretchable strain sensor for human motion detection. *ACS Appl Mater Inter*. 2016;8(37):24837–43.
28. Yan T, Wang Z, Wang YQ, Pan ZJ. Carbon/graphene composite nanofiber yarns for highly sensitive strain sensors. *Mater Design*. 2018;143:214–23.
29. Yan T, Zhou H, Niu HT, Shao H, Wang HX, Pan ZJ, et al. Highly sensitive detection of subtle movement using a flexible strain sensor from helically wrapped carbon yarns. *J Mater Chem C*. 2019;7(32):10049–58.
30. Ye C, Ren J, Wang YL, Zhang WW, Qian C, Han J, et al. Design and fabrication of silk templated electronic yarns and applications in multifunctional textiles. *Matter-Us*. 2019;1(5):1411–25.
31. Zhao HB, Zhang YY, Bradford PD, Zhou QA, Jia QX, Yuan FG, et al. Carbon nanotube yarn strain sensors. *Nanotechnology*. 2010;21(30):305502.
32. Shang YY, He XD, Li YB, Zhang LH, Li Z, Ji CY, et al. Super-stretchable spring-like carbon nanotube ropes. *Adv Mater*. 2012;24(21):2896–900.
33. Shang YY, Li YB, He XD, Du SY, Zhang LH, Shi EZ, et al. Highly twisted double-helix carbon nanotube yarns. *ACS Nano*. 2013;7(2):1446–53.
34. Li YB, Shang YY, He XD, Peng QY, Du SY, Shi EZ, et al. Overtwisted, resolvable carbon nanotube yarn entanglement as strain sensors and rotational actuators. *ACS Nano*. 2013;7(9):8128–35.

35. Ren JS, Wang CX, Zhang X, Carey T, Chen KL, Yin YJ, et al. Environmentally-friendly conductive cotton fabric as flexible strain sensor based on hot press reduced graphene oxide. *Carbon*. 2017;111:622–30.
36. Momin MA, Rahman MJ, Mieno T. Development of compact load cell using multiwall carbon nanotube/cotton composites and its application to human health and activity monitoring. *J Nano Mater*. 2019;2019:1–15.
37. Souri H, Bhattacharyya D. Highly sensitive, stretchable and wearable strain sensors using fragmented conductive cotton fabric. *J Mater Chem C*. 2018;6(39):10524–31.
38. Cao MH, Wang MQ, Li L, Qiu HW, Padhiar MA, Yang Z. Wearable rGO-Ag NW@cotton fiber piezoresistive sensor based on the fast charge transport channel provided by Ag nanowire. *Nano Energy*. 2018;50:528–35.
39. Lopez-Blanco R, Velasco MA, Mendez-Guerrero A, Romero JP, del Castillo MD, Serrano JI, et al. Smartwatch for the analysis of rest tremor in patients with Parkinson's disease. *J Neurol Sci*. 2019;401:37–42.
40. Hui ZY, Chen RY, Chang J, Gong YJ, Zhang XW, Xu H, et al. Solution-processed sensing textiles with adjustable sensitivity and linear detection range enabled by twisting structure. *ACS Appl Mater Inter*. 2020;12(10):12155–64.
41. Liu R, Li JM, Li M, Zhang QH, Shi GY, Li YG, et al. MXene-coated air-permeable pressure-sensing fabric for smart wear. *ACS Appl Mater Inter*. 2020;12(41):46446–54.
42. Li YC, Zheng CR, Liu S, Huang L, Fang TS, Li JXZ, et al. Smart glove integrated with tunable MWNTs/PDMS fibers made of a one-step extrusion method for finger dexterity, gesture, and temperature recognition. *ACS Appl Mater Inter*. 2020;12(21):23764–73.

7 Flexible Strain Sensors Based on Sponges of Conductive Polymer Composites

7.1 INTRODUCTION

There are several reports on the preparation of sponge-based strain sensors in the literature as they have good elasticity and excellent compression properties. This chapter reviews the composition, fabrication techniques, and future applications of sponge-based strain sensors. In particular, we have highlighted our work in preparing sponge-based strain sensors using template methods (including carbonization and coating). Moreover, the characteristics and mechanism of the resistance change of the sponge-based piezoresistive sensors under different strains are described in detail. In summary, sponge-based strain sensors show broad application potential in wearable electronic equipment, human–computer interaction, intelligent robots, electronic skin, and other fields.

7.2 TYPES OF SPONGE-BASED STRAIN SENSORS

Conductive sponges have been widely used in preparing wearable electronic devices [1] thanks to their good elasticity, excellent compression performance, and good electrical conductivity. 3D conductive sponges can be categorized into four types based on their structures and compositions: (i) neat conductive sponges, (ii) conductive sponges impregnated with elastomer, (iii) composite conductive sponges, and (iv) conductive material-coated sponges. Subsequently, several conductive sponges and their preparation process are briefly introduced [2].

7.2.1 Neat Conductive Sponge

Neat conductive sponges have a three-dimensional porous structure assembled only from conductive materials [3]. Currently, nanomaterials widely used in the preparation of neat conductive sponges include graphene nanosheets (GNPs), reduced graphene oxide (rGO) [4], multi-walled carbon nanotubes (MWCNTs), MXene, silver nanowires (AgNWs), and silver nanoparticles (AgNPs) [5]. Although these pure conductive sponges have ultra-low density, high conductivity, and good thermal and chemical stability, they may have limited sensitivity and compressive strain range due to their inherent mechanical rigidity and brittleness. The network structure of these neat conductive sponges is vulnerable to irreversible damage during compression, resulting in severe/irreversible changes in plastic deformation and conductivity. This drawback can be overcome by impregnating the neat conductive sponge with an elastic polymer, referred to as "elastomer impregnated conductive sponge".

7.2.2 Conductive Sponge Impregnated with Elastomer

Elastomer-impregnated conductive sponge is coated/impregnated on the neat conductive sponge surface to improve the compression stability and robustness of the sponge [6]. Currently, thermoplastic polyurethane, silicone rubber, polydimethylsiloxane, and other elastomers have been prevented from being used to prepare elastomer-impregnated conductive sponge. However, the insulating elastomer impregnated on the surface of a neat conductive sponge reduces the conductivity and increases the density of the sponge.

7.2.3 Composite Conductive Sponge

Composite conductive sponges are prepared by foaming or freeze-drying after blending conductive materials with polymers. The conductivity and sensitivity are easily adjusted by changing the number of conductive components in the composite conductive sponge. A thorny problem of the composite conductive sponge is the uneven dispersion of conductive materials and/or polymers in the preparation process. In addition, foaming such composites is challenging due to the increased polymer viscosity because of the filler content [7].

7.2.4 Conductive Material-Coated Sponge

Conductive sponges are made of conductive materials coated on non-conductive polymer sponges by coating, impregnation, magnetron sputtering, or in situ polymerization [8–10]. The affinity and interface interaction between the conductive material and the sponge substrate plays a key role in the conductivity and cycle stability of the composite conductive sponge [11].

7.3 METHODS FOR FABRICATION OF SPONGE-BASED STRAIN SENSORS

7.3.1 Supercritical Foaming Technology

Supercritical foaming molding is both physical and microcellular foaming molding technology. During the injection molding, extrusion, and blow molding process, the supercritical carbon dioxide (ScCO$_2$), the supercritical nitrogen, or other supercritical gases are first injected into a special plasticizing device so that the gas and molten raw materials are thoroughly and evenly mixed/diffused to form a single-phase mixed sol [12]. Then the sol is introduced into the mold cavity or extrusion die, where the sol generates a large pressure drop, and the gas is precipitated to form a large number of bubble cores. In the subsequent cooling molding process, the bubble nucleus inside

Flexible Sponge Strain Sensors

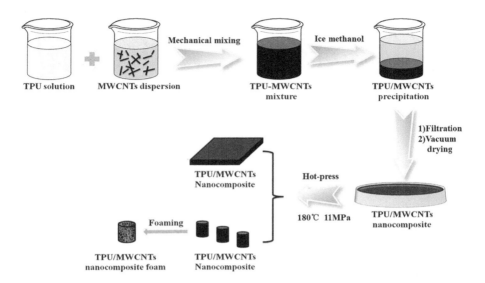

FIGURE 7.1 Preparation process of TPU/MWCNTs composite and sponges. TPU, thermoplastic polyurethane; MWCNTs, multi-walled carbon nanotubes.

the sol forms and grows, finally affording the microcellular foamed plastic products. As shown in Figure 7.1, Fei et al. [13] prepared a piezoresistive sponge based on thermoplastic polyurethane (TPU)/MWCNT composite by solution blending and ScCO$_2$ foaming process, which has high strength, flexibility, and stable cycle. However, due to the high filler content, the viscosity of polymer melting increases, which inhibits the growth of cells and destroys the foaming ability of the polymer.

7.3.2 Chemical Vapor Deposition (CVD)

Deposition of a conductive layer on the porous polymer sponge by CVD can maintain the ideal porous structure of the sponge and provide the sponge with good mechanical and electrical properties [14]. As depicted in Figure 7.2, Xu et al. [15] synthesized three-dimensional graphene (GF)/polydimethylsiloxane (PDMS)-poly(ethylene terephthalate) (PET) sponge by CVD method. The resistance of the GF/PDMS-PET sponge increases with the increasing bending curvature. However, high cost and complex processing technology limit the wide application of the CVD method in preparing sponge-based strain sensors [16].

7.3.3 Freeze-Drying Method

The freeze-drying method is a new method to prepare aerogels with nano-three-dimensional network structure by directly sublimating solvent in polymer solution [17]. A novel flexible piezoresistive sensor was prepared by Huang et al. [18], which was based on porous polyaniline (PANI)/bacterial cellulose (BC)/chitosan (CH) aerogel by simple freeze-drying technology (Figure 7.3). PANI/BC/CH aerogels

130   Carbon-Based Conductive Polymer Composites

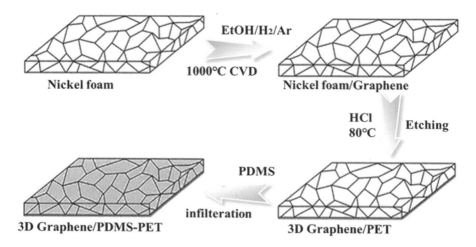

FIGURE 7.2 Schematic presentation of the fabrication of the 3D GF/PDMS-PET composite. GF, graphene; PDMS, polydimethylsiloxane; PET, poly(ethylene terephthalate).

FIGURE 7.3 Digital photographs of (a) BC aerogel, (b) BC/CH aerogel, and (c) PANI/BC/CH aerogel. Various magnification SEM images of (d,g) BC aerogel, (e,h) BC/CH aerogel, and (f,i) PANI/BC/CH aerogel. BC, bacterial cellulose; CH, chitosan; PANI, polyaniline; SEM, scanning electron microscope.

have low density, good electrical conductivity, and wide pressure identification ability. Nevertheless, the wide application of aerogels prepared by freeze-drying technology is limited in flexible electronic devices due to their high cost restricting their mass production and poor mechanical properties [19,20].

7.3.4 TEMPLATE METHOD

7.3.4.1 Carbonization of Template Sponge

High-temperature pyrolysis or carbonization is the most straightforward method to produce carbon material. Phenol formaldehyde foam (PFF) is a common thermosetting resin widely used in the construction industry for heat and sound insulation applications. The preparation of a flexible multipurpose piezoresistive strain sensor reported by Wang et al. [21] is illustrated in Figure 7.4a, which relies on carbonized PFF as a template and encapsulation with PDMS. First, commercially purchased PFF was converted into carbonized PFF (cPFF) with adjustable conductivity by simple high-temperature pyrolysis (Figure 7.4b). Then, the cPFF was encapsulated with PDMS as elastomer to prepare the piezoresistive strain sensor. The sensor exhibited high sensitivity, stable current response, and excellent durability over many loading cycles. Furthermore, large-scale (finger bending and neck movement) and small-scale (facial micro-expression and phonation) human motions were successfully monitored, thereby demonstrating the potential of the proposed sensor in smart wearable devices. Wang et al. [22] reported another high-performance flexible piezoresistive sensor based on carbonized poly-benzoxazine (cPBa) foam and PDMS using the same strategy. Moreover, they found that due to the disconnection/reconnection mechanism of carbon debris, the cPBa/PDMS strain sensor exhibits negative and positive pressure resistance effects under different pressures. As evident from Figure 7.5, initial small strain ($\varepsilon < 6\%$) induces the deflection and separation of the carbonized fragments in PDMS, interrupting the direct connection of adjacent carbon fragments, which destroys the conductive network in the sponge. However, due to continuous compression, the direct distance of carbon fragments decreases, which

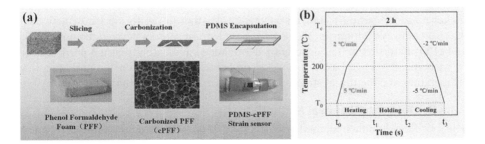

FIGURE 7.4 (a) Fabrication of the PDMS-cPFF multipurpose strain sensor and (b) related carbonization process parameters. PDMS, polydimethylsiloxane; cPFF, carbonized phenol formaldehyde foam.

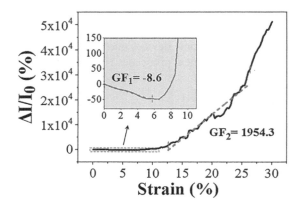

FIGURE 7.5 Relative electrical current change (RCV) as a function of compression strain.

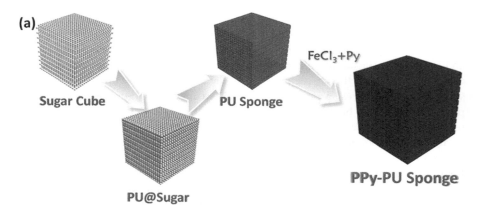

FIGURE 7.6 Schematic illustration of sugar-templated preparation of PPy-PU sponge. PPy, polypyrrole; PU, polyurethane.

makes adjacent carbon fragments reconnect, leading to the formation of new conductive pathways. Moreover, the sponge-based strain sensors derived from the cPBa and cPFF foams exhibited different performances, indicating that the performance of carbonized sponge-based strain sensors strongly depends on the carbonization temperature and the initial structure of the material before carbonization.

7.3.4.2 Template Removal Method

The sacrificial template method is widely used to prepare porous structures, in which the sacrificial template material serves as a temporary framework in assisting the formation of desired 3D structure. Wan et al. [23] utilized the sacrificial sugar template method to prepare a polypyrrole (PPy)-PU conductive sponge-based strain sensor with a unique micro-wrinkle structure (Figure 7.6). These PPy-PU sponges exhibited good mechanical properties and electrical conductivity due to the combination

Flexible Sponge Strain Sensors

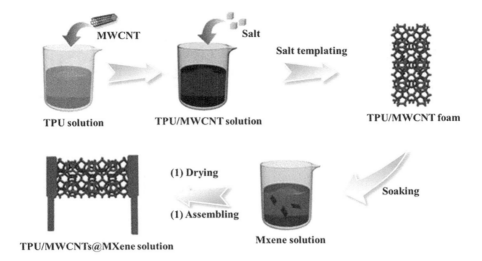

FIGURE 7.7 Scheme of the fabrication process of the TPU/MWCNTs@MXene foam strain sensor. TPU, thermoplastic polyurethane; MWCNTs, multi-walled carbon nanotubes.

of elastomeric PU and conductive PPy, making them good candidates to prepare pressure and strain-sensitive materials. As displayed in Figure 7.7, Wang et al. [24] also reported the fabrication of a porous sponge-based strain sensor composed of $Ti_3C_2T_x$ MXene, MWCNTs, and TPU. MXene sheets were adsorbed on the surface of the MWCNTs/TPU sponge prepared by the sacrificial salt-templating method, thereby further improving the conductivity and sensing performance of sponge-based sensors.

7.3.4.3 Surface Coating of a Template Sponge

The surface coating strategy based on the commercial sponge is a simple, direct, and effective method for preparing conductive sponge-based strain sensors [25]. A layer of conductive material can be deposited on the existing polymer sponges (such as polyurethane and melamine) or as-prepared sponges. These sponge-based strain sensors generally maintain the outstanding compression performance of the pure sponge. Good adhesion between the template sponge and the conductive coating material is the key to ensuring the stability and reproducibility of the conductive sponge-based strain sensors [26]. Zhang and coworkers [27] prepared a flexible and high-performance piezoresistive strain sensor by simple layer-by-layer electrostatic self-assembly of carbon nanoparticles on a polyurethane (PU) sponge (Figure 7.8). The alternate assembly of carbon nanotubes (CNTs) and graphene nanoplatelets (GNPs) afforded a more complete conductive network, and the synergistic effect between CNTs and GNPs significantly improved the sensing performance of the sensor. The current of the sensor decreased with increasing tensile strain, showing a positive piezoresistive effect (Figure 7.9a). However, the current increased when the sponge sensor was compressed, showing a negative piezoresistive effect (Figure 7.9b). During the deformation process of the sponge-based strain sensor, the "destruction"

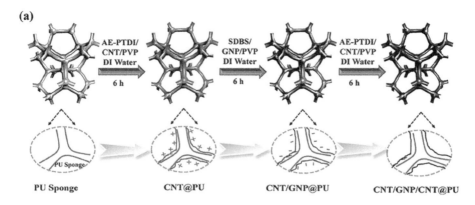

FIGURE 7.8 Schematic diagram for the preparation of the PU sponge-based strain sensors. PU, polyurethane.

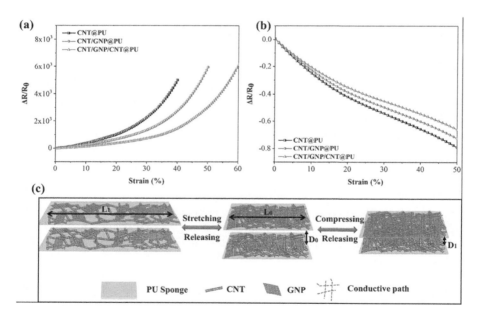

FIGURE 7.9 $\Delta R/R_0$-strain curves for the sponge-based sensors: (a) tensile strain, (b) compressive strain, and (c) schematic diagram for the evolution of conductive paths in the strain sensors under different strains.

and "reconstruction" of the conductive paths proceed simultaneously. However, the destruction of the conductive network dominated during the tensile process, increasing the tunneling distance between adjacent conductive nanoparticles. When the sensors were compressed, more conductive paths were reconstructed, increasing the conductivity (Figure 7.9c).

Flexible Sponge Strain Sensors

7.4 APPLICATION OF SPONGE-BASED STRAIN SENSORS

7.4.1 Wearable Electronic Device

The rapid development of wearable electronic devices puts forward new requirements for the performance and preparation of strain sensors. Meanwhile, wearable electronic devices require each component to have high compressibility to adapt to large-scale mechanical deformation and complex conditions. The sponge-based strain sensor with good elasticity, high compressibility, and excellent electrical conductivity can successfully meet the needs of monitoring pressure changes during most human activities. For example, Song et al. [28] prepared a sponge-based piezoresistive sensor based on CNTs and PDMS (CNT-PDMS), as shown in Figure 7.10,

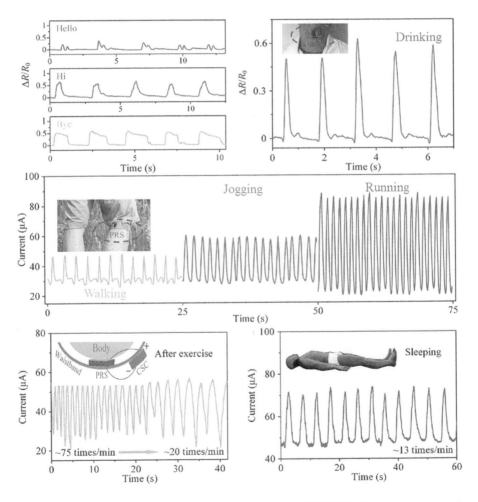

FIGURE 7.10 Electromechanical response phenomenon of CNT-PDMS sponge while monitoring human activities. CNT, carbon nanotube; PDMS, polydimethylsiloxane.

which can be attached to skin or clothing and can detect human actions from speech recognition to respiratory records, showing great potential for application in wearable electronic devices.

7.4.2 Human–Computer Interaction/Intelligent Robot

Traditional wearable devices have single functions and relatively fewer characteristics, which do not meet the needs of modern people for intelligent life. However, the intelligent wearable devices are connected to the internet, smartphones, tablets, and computers for interaction, forming the intelligent wearable device-based Internet of Things (IOT) terminal. For instance, a novel flexible piezoresistive tactile sensor with Ti$_2$C-PDMS sponge as conductive elastomer was fabricated by Sun et al. [29] prepared. This sensor has excellent sensing performance and can realize human–computer interaction by assembling a matrix (Figure 7.11).

7.4.3 Electronic Skin

Electronic skin is a new electronic device that simulates the external stimulation (e.g., pressure, temperature, humidity) of human skin through the integration and feedback of electrical signals. As a kind of flexible tactile bionic sensor, electronic skin has been widely used in human physiological parameters detection, robot tactile perception, etc. Miao et al. [30] designed a new method to combine wrinkle structure with porous sponge structure, achieving a novel, flexible, compressible, and bifunctional sensor based on CNT-PDMS with humidity and pressure sensing functions. The performance of the humidity sensing part can be controlled by ultraviolet (UV) and ozone (UVO) treatment time and CNT concentration, and the sensitivity of the pressure sensing part can be controlled by CNT concentration and grinding time of sugar particles. Besides humidity change and tactile perception functions, the sensor can also monitor human motion and health (Figure 7.12).

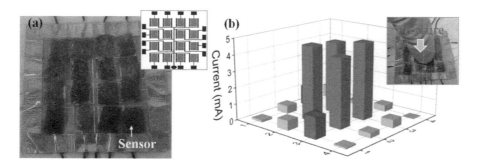

FIGURE 7.11 (a) Photo of the sensor array (Inset: Structural diagram of the sensor array). (b) Response of the sensor array when pressing on the different positions.

Flexible Sponge Strain Sensors 137

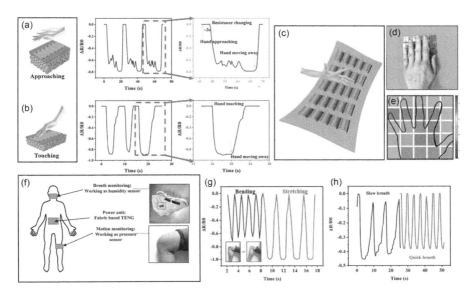

FIGURE 7.12 (a) Humidity sensing for hands approaching. (b) Pressure sensing for hands touching. (c) Schematic illustration of the sensor array. (d, e) Sensor array for sensing force. (f) Sensor fixed on a human knee joint or integrated with a mask works as a flexible health monitor. (g) Different resistance changes between normal high knee holding and too high-level knee holding. (h) Detecting various breathing states: slow breath and quick breath.

7.5 CONCLUSIONS

Because of the good elasticity and excellent compression properties of the sponge, there are several reports on the preparation of sponge-based strain in the literature. The composition, fabrication techniques, and future applications of sponge-based strain sensors are reviewed in this chapter. Our work in preparing sponge-based strain sensors using template methods (including carbonization and coating) is particularly emphasized. The simplest and most convenient method for the large-scale preparation of flexible sponge-based strain sensors is carbonization. The performance of carbonized sponge-based strain sensors significantly depends upon the carbonization temperature and the initial structure of the material before carbonization. Another commonly used method for preparing sponge-based high-performance flexible strain sensors is the layer-by-layer electrostatic self-assembly coating. In this method, nanofillers with different electronegativity can be loaded on the sponge surface through proper surface modification. The coordination between different fillers helps to enhance further the electrical and sensing performance of the sponge-based sensor. Moreover, the characteristics and mechanism of the resistance change of the sponge-based piezoresistive sensor under different strains are described in detail. In summary, sponge-based strain sensors show broad application potential in wearable electronic equipment, human–computer interaction, intelligent robots, electronic skin, and other fields.

REFERENCES

1. Sang Z, Ke K, Manas-Zloczower I. Design strategy for porous composites aimed at pressure sensor application. *Small*. 2019;15(45):e1903487.
2. Ding Y, Xu T, Onyilagha O, Fong H, Zhu Z. Recent advances in flexible and wearable pressure sensors based on piezoresistive 3D monolithic conductive sponges. *ACS Appl Mater Interfaces*. 2019;11(7):6685–704.
3. Sun S, Liu Y, Chang X, Jiang Y, Wang D, Tang C, et al. A wearable, waterproof, and highly sensitive strain sensor based on three-dimensional graphene/carbon black/Ni sponge for wirelessly monitoring human motions. *J Mater Chem C*. 2020;8(6):2074–85.
4. Boland CS, Khan U, Binions M, Barwich S, Boland JB, Weaire D, et al. Graphene-coated polymer foams as tuneable impact sensors. *Nanoscale*. 2018;10(11):5366–75.
5. Samad YA, Li Y, Schiffer A, Alhassan SM, Liao K. Graphene foam developed with a novel two-step technique for low and high strains and pressure-sensing applications. *Small*. 2015;11(20):2380–5.
6. Yu XG, Li YQ, Zhu WB, Huang P, Wang TT, Hu N, et al. A wearable strain sensor based on a carbonized nano-sponge/silicone composite for human motion detection. *Nanoscale*. 2017;9(20):6680–5.
7. Zhang BX, Hou ZL, Yan W, Zhao QL, Zhan KT. Multi-dimensional flexible reduced graphene oxide/polymer sponges for multiple forms of strain sensors. *Carbon*. 2017;125:199–206.
8. Wu X, Han Y, Zhang X, Zhou Z, Lu C. Large-area compliant, low-cost, and versatile pressure-sensing platform based on microcrack-designed carbon black@polyurethane sponge for human-machine interfacing. *Adv Funct Mater*. 2016;26(34):6246–56.
9. Zhao L, Qiang F, Dai SW, Shen SC, Huang YZ, Huang, et al. Construction of sandwich-like porous structure of graphene-coated foam composites for ultrasensitive and flexible pressure sensors. *Nanoscale*. 2019;11(21):10229–38.
10. Zheng Q, Liu X, Xu H, Cheung MS, Choi YW, Huang HC, et al. Sliced graphene foam films for dual-functional wearable strain sensors and switches. *Nanoscale Horiz*. 2018;3(1):35–44.
11. Xiang D, Zhang Z, Han Z, Zhang X, Zhou Z, Zhang J, et al. Effects of non-covalent interactions on the properties of 3D printed flexible piezoresistive strain sensors of conductive polymer composites. *Compos Interfaces*. 2020;28(6):577–91.
12. Li J, Li W, Huang W, Zhang G, Sun R, Wong CP. Fabrication of highly reinforced and compressible graphene/carbon nanotube hybrid foams via a facile self-assembly process for application as strain sensors and beyond. *J Mater Chem C*. 2017;5(10):2723–30.
13. Fei Y, Chen F, Fang W, Xu L, Ruan S, Liu X, et al. High-strength, flexible and cycling-stable piezo-resistive polymeric foams derived from thermoplastic polyurethane and multi-wall carbon nanotubes. *Compos Part B Eng*. 2020;199:108279.
14. Zhao X, Xu L, Chen Q, Peng Q, Yang M, Zhao W, et al. Highly conductive multifunctional rGO/CNT hybrid sponge for electromagnetic wave shielding and strain sensor. *Adv Mater Technol*. 2019;4(9):1900443.
15. Xu R, Lu Y, Jiang C, Chen J, Mao P, Gao G, et al. Facile fabrication of three-dimensional graphene foam/poly(dimethylsiloxane) composites and their potential application as strain sensor. *ACS Appl Mater Interfaces*. 2014; 4(16): 13455-60.
16. Jeong YR, Park H, Jin SW, Hong SY, Lee SS, Ha JS. Highly stretchable and sensitive strain sensors using fragmentized graphene foam. *Adv Funct Mater*. 2015;25(27):4228–36.
17. Zeng Z, Seyed Shahabadi SI, Che B, Zhang Y, Zhao C, Lu X. Highly stretchable, sensitive strain sensors with a wide linear sensing region based on compressed anisotropic graphene foam/polymer nanocomposites. *Nanoscale*. 2017;9(44):17396–404.
18. Huang J, Li D, Zhao M, Ke H, Mensah A, Lv P, et al. Flexible electrically conductive biomass-based aerogels for piezoresistive pressure/strain sensors. *Chem Eng J*. 2019;373:1357–66.

19. Tay RY, Li H, Lin J, Wang H, Lim JSK, Chen S, et al. Lightweight, superelastic boron nitride/polydimethylsiloxane foam as air dielectric substitute for multifunctional capacitive sensor applications. *Adv Funct Mater.* 2020;30(10):1–8.
20. Qin Y, Peng Q, Ding Y, Lin Z, Wang C, Li Y, et al. Lightweight, superelastic, and mechanically flexible graphene/polyimide nanocomposite foam for strain sensor application. *ACS Nano.* 2015;9(9):8933–41.
21. Wang L, Xiang D, Harkin-Jones E, Zhang X, Li Y, Zheng Y, et al. A flexible and multipurpose piezoresistive strain sensor based on carbonized phenol formaldehyde foam for human motion monitoring. *Macromol Mater Eng.* 2019;304(12):1900492.
22. Wang L, Xiang D, Zhu W, Zhao C, Li Y, Han H, et al. Flexible piezoresistive strain sensor with high sensitivity based on carbonised waste thermosetting resin. *Plast Rubber Compos.* 2020;49(7):300–6.
23. Wan Y, Qin N, Wang Y, Zhao Q, Wang Q, Yuan P, et al. Sugar-templated conductive polyurethane-polypyrrole sponges for wide-range force sensing. *Chem Eng J.* 2020;383:123103.
24. Wang H, Zhou R, Li D, Zhang L, Ren G, Wang L, et al. High-performance foam-shaped strain sensor based on carbon nanotubes and $Ti_3C_2T_x$ MXene for the monitoring of human activities. *ACS Nano.* 2021;15(6):9690–700.
25. Zhang S, Liu H, Yang S, Shi X, Zhang D, Shan C, et al. Ultrasensitive and highly compressible piezoresistive sensor based on polyurethane sponge coated with a cracked cellulose nanofibril/silver nanowire layer. *ACS Appl Mater Interfaces.* 2019;11(11):10922–32.
26. Ma Z, Wei A, Ma J, Shao L, Jiang H, Dong D, et al. Lightweight, compressible and electrically conductive polyurethane sponges coated with synergistic multi-walled carbon nanotubes and graphene for piezoresistive sensors. *Nanoscale.* 2018;10(15):7116–26.
27. Zhang X, Xiang D, Zhu W, Zheng Y, Harkin-Jones E, Wang P, et al. Flexible and high-performance piezoresistive strain sensors based on carbon nanoparticles@polyurethane sponges. *Compos Sci Technol.* 2020;200:108437.
28. Song Y, Chen H, Su Z, Chen X, Miao L, Zhang J, et al. Highly compressible integrated supercapacitor-piezoresistance-sensor system with CNT-PDMS sponge for health monitoring. *Small.* 2017;13(39):1702091.
29. Sun P, Wu D, Liu C. High-sensitivity tactile sensor based on Ti_2C-PDMS sponge for wireless human–computer interaction. *Nanotechnology.* 2021;32:295506.
30. Miao L, Wan J, Song Y, Guo H, Chen H, Cheng X, et al. Skin-inspired humidity and pressure sensor with a wrinkle-on-sponge structure. *ACS Appl Mater Interfaces.* 2019;11(42):39219–27.

8 3D-Printed Flexible Strain Sensors of Conductive Polymer Composites

8.1 INTRODUCTION

3D printing (namely additive manufacturing) is a rapid prototyping technology that allows complex structures to be printed without the need for any mold tooling [1]. This reduces capital costs and increases lead times as well as allowing the manufacture of structures that cannot be made by any other method. Currently, 3D printing has multiple classifications to meet different materials, such as metals and polymers by selective laser sintering (SLS), photocurable materials by stereolithography (SLA), direct ink writing (DIW), and thermoplastics by fused deposition modeling (FDM) [1–8]. In recent years, a lot of work has been carried out to develop the 3D printing technology for polymer composites.

More attention has been paid to the design and preparation of flexible strain sensors using the characteristics of low cost, high precision, flexibility, and designability of 3D printing. This chapter reviews the composition, preparation methods, structural characteristics, and future application prospects of 3D printing strain sensors. In particular, we focus on some of our work in the preparation of flexible strain sensors using FDM method and introduce non-covalent bond modification or coordination effect between different dimensional direct nanofillers to further improve the overall performance of the sensor. In summary, 3D-printed strain sensor shows broad application potential in wearable electronic equipment, human–computer interaction, intelligent robots, electronic skin, and other fields.

8.2 PREPARATION OF 3D-PRINTED STRAIN SENSORS

3D printing technology generates 3D entities by layer-by-layer processing, overlaying, and adding materials. First, the three-dimensional model of the required components is designed using the computer [1]. Then, according to the process requirements, the model is dispersed into a series of orderly units according to certain rules. Subsequently, the NC code is automatically generated when the system runs, after inputting the processing parameters according to the contour information of each slice. Finally, a series of slices are formed and connected automatically, and finally a three-dimensional physical entity is obtained. From the current report, the manufacturing of 3D printing strain sensor can be divided into digital photocuring (DLP), DIW, FDM, and SLS [9].

DOI: 10.1201/9781003218661-8

8.2.1 DIW-Based 3D-Printed Strain Sensors

DIW 3D printing technology can be used to prepare materials with various materials and properties, and its application fields are very wide, including electronics, structural materials, tissue engineering, and soft robots. There are many types of ink used in this technology, such as conductive adhesive, elastomer, and hydrogel [10]. These inks have rheological properties (such as viscoelasticity, shear thinning, and yield stress), which are helpful for the implementation of 3D printing process. In the DIW process, viscoelastic ink is extruded from the nozzle of 3D printer to form fiber. As the nozzle moves, it can be deposited in a specific pattern. As shown in Figure 8.1, Valentine et al. [11] prepared a new type of high-performance electronic skin by DIW using Ag-thermoplastic polyurethane (TPU) conductive ink. By developing

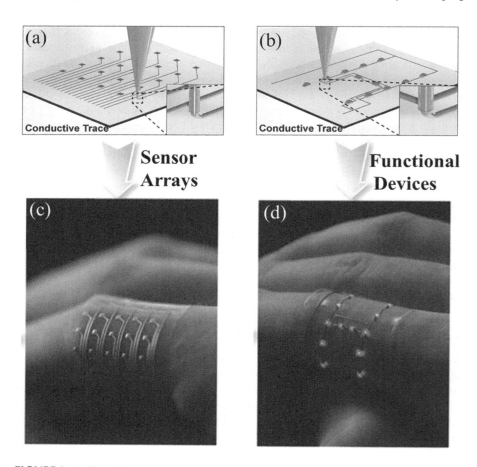

FIGURE 8.1 Hybrid 3D printing platform for soft electronics. Schematic images of (a) DIW of a TPU matrix for the device body and (b) DIW of conductive traces to interconnect the surface mounted LEDs placed in the form of a soft, stretchable "H" LED array; (c) Image of representative example of soft sensor array; (d) Image of a functional LED array wrapped around a human finger.

3D-Printed Flexible Strain Sensors

novel insulating and conductive inks, functional strain and pressure sensors were fabricated and characterized. This hybrid method enables surface mount electrical components of arbitrary shapes and sizes to be readily integrated onto printed soft wearable circuits.

8.2.2 SLA-Based 3D-Printed Strain Sensors

SLA is under the control of digital signal, and the liquid photosensitive resin in the working chamber of the nozzle forms a droplet in an instant, and the nozzle is sprayed to the specified position under pressure. Then the photosensitive resin is cured by ultraviolet light, and the cured resin is stacked layer by layer to obtain the formed parts. SLA material is composed of photocuring solid material and supporting material. The supporting material can be divided into phase change wax supporting material and photocuring supporting material according to different curing methods. From Figure 8.2, Pan et al. [12] reported the SLA 3D printing of tough, highly solvated, and antifouling hybrid hydrogels for potential uses in biomedical, smart materials, and sensor applications. The zwitterionic hybrid hydrogels (Z-gels) could be tuned over a large range of ultimate strains, while also demonstrating a high resilience under cyclic tensile loading.

8.2.3 SLS-Based 3D Printed Strain Sensors

SLS process uses powder materials. Laser selectively scans the powder under the control of the computer to make it melt and then cools and solidifies to achieve the sintering bonding of the material so that the material is stacked layer by layer to achieve molding. Wei et al. [13] reported a novel, simple, and low-cost SLS strategy to produce a soft conductive film by sintering a mixture of nylon and graphite powder with a variety of detection capabilities for touch, stretch, and bending (as shown in Figure 8.3). The soft materials present the combination of high sensitivity and good repeatability. An elastic robotic arm was constructed, which has a nylon-graphite sensor on the surface that simultaneously enables active haptic and proprioceptive sensing for different objects with various weight, shape, and roughness.

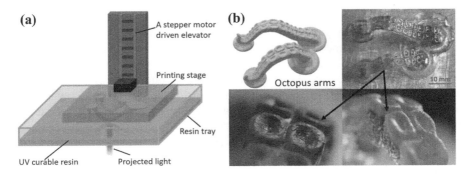

FIGURE 8.2 (a) Schematic diagram of the projection-stereolithography; (b) Rapid fabrication of octopus arms with 50 µm resolution at a speed of 14 seconds per layer using SLA.

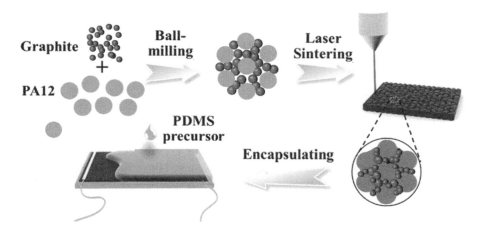

FIGURE 8.3 Schematic illustration of the fabrication process of the CPC-based sensor. CPC, conductive polymer composite.

8.2.4 FDM-Based 3D-Printed Strain Sensors

FDM is a polymer 3D printing technology that combines computer-aided design with thermoplastic polymer materials and is formed by layer-by-layer overlay of melt filaments [14,15]. Compared with other 3D printing methods, FDM has the characteristics of low cost, convenience, and flexibility, and the extruded filaments are suitable for large-scale production and long-term storage. From Figure 8.4, Xiang et al. [16] fabricated highly flexible strain sensors based on carbon nanotube (CNT)/TPU, graphene nanoplatelet (GNP)/TPU, and CNT/GNP/TPU nanocomposites by FDM 3D printing. The dispersion, printability, as well as the electrical, tensile, and sensing properties of the printed composites were systematically investigated. Due to the synergistic effect of CNTs and GNPs, the improved dispersion of CNTs in the TPU matrix was obtained, and the electrical and tensile properties of the 3D-printed sensor were significantly enhanced. To analyze the mechanism of strain sensing, modeling based on tunnel theory was performed. Xiang et al. [17] prepared CNT/TPU nanocomposites by solution mixing and extruded into filaments using a single-screw extruder for fabricating flexible strain sensors by FDM. Since π-π non-covalent interactions can be generated between CNTs and 1-pyrenecarboxylic acid (PCA) through a conjugation effect and hydrogen bonding can also be formed between the carboxyl groups of PCA and the carbonyl and amide groups of TPU, PCA was used to non-covalently modify the nanotubes in order to improve their dispersion and alter polymer–nanofiller interactions without destruction of the intrinsic structure of the CNTs. Xiang et al. [18] prepared ternary CNT/AgNP/TPU nanocomposites by solution mixing and extruded into filaments using a single-screw extruder for subsequent printing into flexible strain sensors by FDM. The AgNPs in that study prevent the intact interaction between neighboring CNTs as well as tailoring the conductive pathways in the composite. In addition, the sensing properties such as strain detection range, sensitivity, linearity, response time, resistance responsiveness, stability,

3D-Printed Flexible Strain Sensors

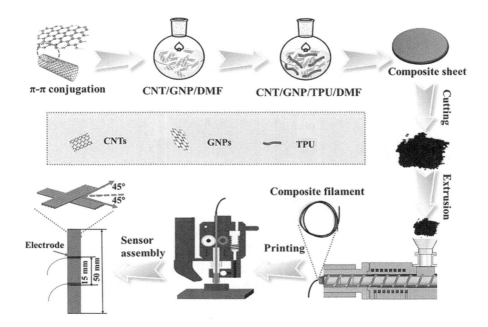

FIGURE 8.4 Schematic diagram of the 3D printing of the flexible strain sensor.

and durability under different conditions are equally significant for conductive polymer composite (CPC)-based strain sensors.

8.3 CONDUCTIVE MATERIALS FOR 3D-PRINTED STRAIN SENSORS

8.3.1 CARBON MATERIALS

Carbon-based materials have the advantages of strong conductivity, good stability, and low price. They are widely used to prepare tensile conductive composites and have great application potential in the fields of tensile and wearable electronic equipment, which has attracted close attention of researchers [5]. Currently, the carbon-based materials widely used in the preparation of flexible sensors include GNPs, reduced graphene oxide (rGO), multi-walled carbon nanotubes (MWCNTs), and carbon black (CB) [19]. As shown in Figure 8.5, Xiang et al. [20] fabricated a high-performance fiber strain sensor by constructing a double percolated structure, consisting of CNT/TPU continuous phase and styrene butadiene styrene (SBS) phase, incompatible with TPU (CNT/TPU@SBS). Compared with other similar fiber strain sensor systems without double percolated structure, the CNT/TPU@SBS sensor achieves a lower percolation threshold (0.38 wt%) and higher electrical conductivity. The conductivity of 1%-CNT/TPU@SBS (4.12×10^{-3} S m^{-1}) is two orders of magnitude higher than that of 1%-CNT/TPU (3.17×10^{-5} S m^{-1}) at the same CNT loading of 1 wt%. Due to double percolated structure, the 1%-CNT/TPU@SBS sensor exhibits a wide strain detection range (0.2%–100%) and an ultra-high sensitivity (maximum gauge factor

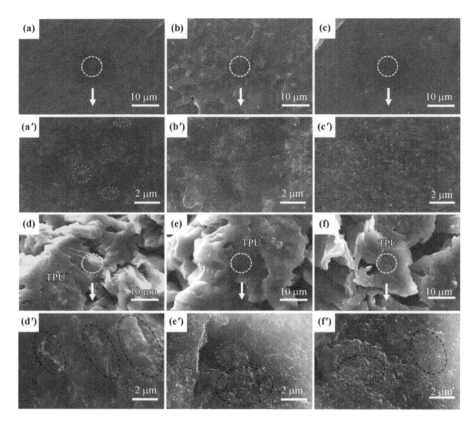

FIGURE 8.5 The morphologies of uniformly dispersed network of (a–c, a'–c') CNT/TPU and (d–f, d'–f') double percolated network of CNT/TPU@SBS with different CNT loadings after etching. CNT, carbon nanotube; TPU, thermoplastic polyurethane.

(GF) is 32,411 at 100% strain). Besides, the 1%-CNT/TPU@SBS sensor shows a high linearity ($R^2 = 0.97$) at 0%–20% strain, relatively fast response time (214 ms), and stability (500 loading/unloading cycles). Xiang et al. [21] prepared flexible piezoresistive strain sensors of CNT/TPU CPCs by FDM. The printed sensors exhibited a typical negative piezoresistive effect under compressive strain. The effects of non-covalent modification and loading of CNTs on the mechanical and piezoresistive behavior of sensors were investigated. The introduction of PCA) increased the dispersibility of CNTs and nanofiller–polymer interfacial interactions, resulting in the excellent sensing performance. From Figure 8.6, Liu et al. [14] printed a novel, highly flexible, and electrically resistive-type strain sensor with a special three-dimensional conductive network using a composite of conductive graphene pellets and flexible TPU pellets. The strain sensor was then printed onto a glass substrate using the composite filament with an FDM-based 3D printer. Chen et al. [8] demonstrated the FDM 3D printing nanocomposites of TPU/poly(lactic acid)/graphene oxide (TPU/PLA/GO) and their potential applications as biocompatibility materials. From Figure 8.7, the

3D-Printed Flexible Strain Sensors   147

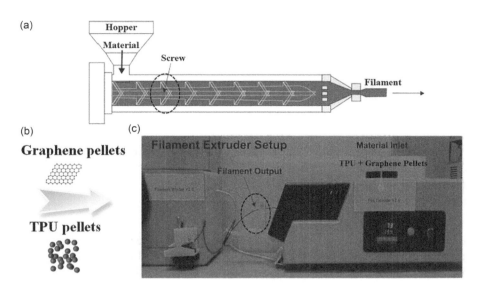

FIGURE 8.6 (a) Illustration of the filament extrusion mechanism. (b and c) Diagram of the graphene pellets, TPU pellets, and commercial filament extrusion setup. TPU, thermoplastic polyurethane.

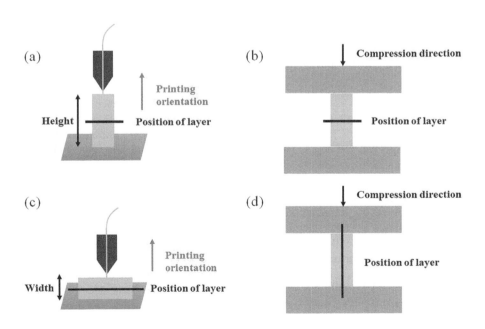

FIGURE 8.7 (a) Schemes of standing specimen 3D printing. (b) Illustration of S compression testing. (c) Schemes of lying specimen 3D printing. (d) Illustration of L compression testing.

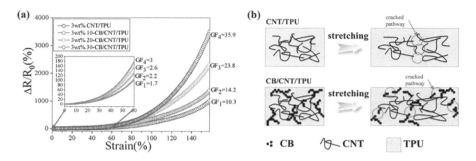

FIGURE 8.8 (a) Relative resistance changes of the sensors during stretching. (b) Schematic diagram of stretched conductive networks.

experiment shows that the content of graphene and the degree of orientation in the printing process have a significant impact on the mechanical and electrical properties of the sensor. From Figure 8.8, Zhang et al. [22] fabricated flexible piezoresistive strain sensors based on CNT- and CB-filled TPU composites (CNT/TPU and CB/CNT/TPU). The conductive network constructed by CNTs in the TPU matrix was destroyed by stretching, and the sensor exhibited a typical positive piezoresistive effect during the stretching process. The effect of CB addition on the performance of the sensor was also studied. The results showed that CB formed some new conductive paths in the unconnected positions of the CNT network, which enhanced the conductivity of the composites, and the conductivity of the composites increased with the increase in CB content.

8.3.2 Metal Material/MXene

Metal fillers (silver nanowires, silver nanoparticles, silver nanosheets, and gold nanorods) and MXene as conductive fillers with excellent performance are one of the important raw materials for preparing flexible strain sensors by 3D printing [4,23]. As shown in Figure 8.9, Britton et al. [24] prepared an electronic ink based on silver nanowires (AgNWs) and copolymer-x-pentadecalactone-co-e-decalactone (PDL) and printed it as an environmentally sustainable strain sensor, which proved that modern additive manufacturing technology could use direct writing process to manufacture microscale finger electrode design. Kwon et al. [25] prepared a multiaxial piezoresistive sensor based on graphene nanoplatelets (GNPs), silver nanoparticles (AgNPs), and polyurethane by DIW (Figure 8.10). Based on the synergistic effect between GNPs and AgNPs, the 3D-printed GNPs/AgNPs nanocomposite sensor has demonstrated excellent performance as a highly sensitive multi-axial piezoresistive sensor. From Figure 8.11, Cao et al. [26] prepared a flexible intelligent fabric fiber based on MXene and TEMPO (2,2,6,6-tetramethylpiperidin-1-yloxy radical) ink by ink-jet printing technology. The fiber and textiles have excellent response to various external stimuli (electrical/photonic/mechanical). TEMPO-oxidized cellulose nanofibers (TOCNFs)/Ti_3C_2 in hybrid inks self-assemble to fibers with an aligned structure in ethanol, mimicking the features of the natural structures of plant fibers.

3D-Printed Flexible Strain Sensors

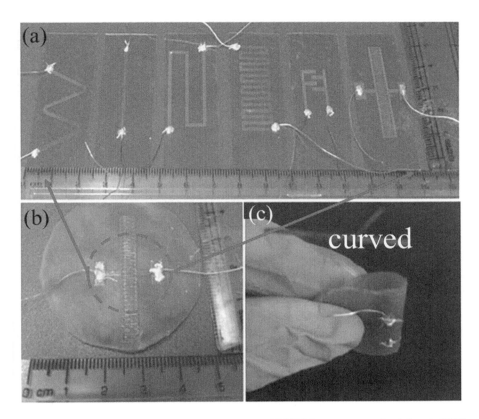

FIGURE 8.9 Extrusion printing of conductive AgNW/PDL nanocomposite inks. AgNW/PDL inks were extruded onto thermanox. PDL, pentadecalactone-*co*-e-decalactone; AgNW, silver nanowires. (a) AgNW/PDL inks were extruded onto thermanox and (b) pristine cast PDL substrates; (c) The flexibility of the extrusion printed biosensors is illustrated.

8.3.3 Conductive Hydrogel

Hydrogel is a kind of water-containing network polymer, which has been widely used in biomedicine, flexible electronics, and other technologies. In many applications, hydrogels combine with other polymers to form a hybrid structure for protecting, enhancing, or adding new functions to hydrogel structures [27], such as elastic biomedical equipment with hydrophilic and lubricating surface based on hydrogel, flexible electronic equipment based on hydrogel with elastomer anti-dehydration coating, and hydrogel composites reinforced by elastomer fiber. From Figure 8.12, Liu et al. [28] selected polysaccharide-κ-carrageenan, prepared κ-carrageenan/polyacrylamide (PAAm) double network (DN) hydrogel with excellent mechanical properties and recyclable by 3D printing technology through physical and thermal reversible network, and further studied the toughening mechanism, self-healing properties, 3D printing ability, and strain sensitivity of κ-carrageenan/PAAm DN hydrogel. More importantly, the warm pregel solution of κ-carrageenan/AAm was successfully used as an ink of a 3D printer to print

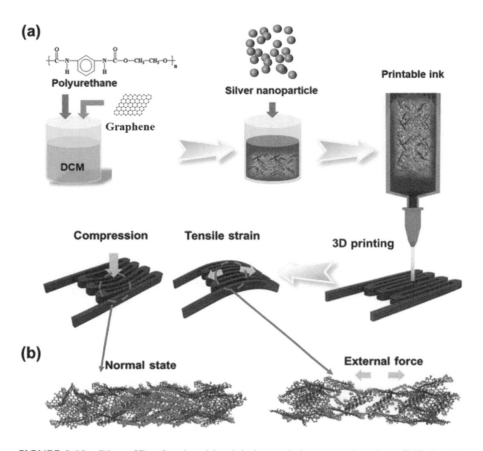

FIGURE 8.10 Direct 3D-printed multi-axial piezoresistive sensor based on GNPs/AgNPs nanocomposite. (a) Schematic illustration of the fabrication of GNPs/AgNPs nanocomposite and the direct 3DP of multi-axial piezoresistive sensor by liquid deposition method. (b) Schematic illustration of the sensing mechanism. GNPs, graphene nanoplatelets; AgNPs, silver nanoparticles.

the hydrogel into complex 3D structures, and the printed DN hydrogel samples demonstrated the high mechanical strength after UV exposure. From Figure 8.13, Williams et al. [29] proposed the use of dendricolloids to enhance the mechanical strength of hydrogels, improving the printing performance of homocomposite hydrogels (HHGs).

8.4 ARCHITECTURAL DESIGN FOR 3D-PRINTED STRAIN SENSORS

8.4.1 Micro-Nano Porous Structure

Compared with other methods for preparing porous materials (such as foaming technology and freeze drying method), 3D printing technology has high precision, low cost, and can rapidly and accurately prepare strain sensors with porous structures [30].

3D-Printed Flexible Strain Sensors

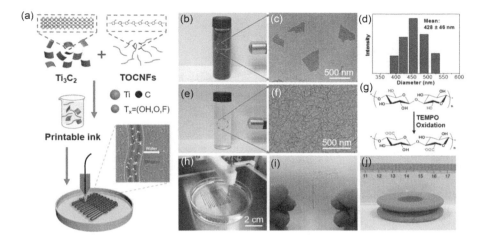

FIGURE 8.11 (a) Schematic illustration of the fabrication of smart TOCNFs/Ti$_3$C$_2$ fibers and textiles. (b) Optical image, (c) TEM image, and (d) diameter distribution of Ti$_3$C$_2$ MXene nanosheets. (e) Optical image and (f) TEM image of TOCNFs. (g) Schematic of the TEMPO oxidation process to prepare TOCNFs. (h) Photograph of the TOCNFs/Ti$_3$C$_2$ fiber during the printing process. (i) Optical image of the TOCNFs/Ti$_3$C$_2$ fibers. (j) A 1 m long TOCNFs/Ti$_3$C$_2$ fiber on a bobbin winder.

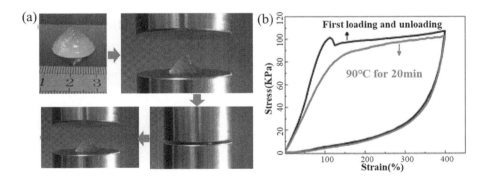

FIGURE 8.12 (a) Photos of 3D-printed κ-carrageenan/PAAm DN hydrogel during mechanical testing. (b) Strain-stress curves of the sample.

With the high freedom of structure design, 3D printing shows great promise to fabricate such complex elastic micro-nano porous structure. From Figure 8.14, Duan et al. [31] prepared 3D porous polydimethylsiloxane (O-PDMS) by 3D printing technology and combined it with CNTs and graphene conductive network to form a high stretchable conductor (OPCG). Two kinds of OPCG were prepared, and the finite element method was used to verify that the OPCG with split structure has high conductivity and excellent retention ability under deformation (Figure 8.15). Wang et al. [32]

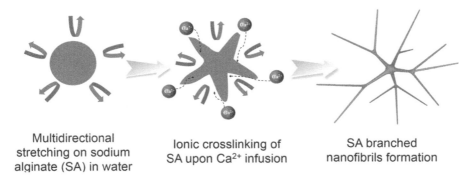

FIGURE 8.13 Schematic of the SA SDC formation process by precipitation in a turbulent medium. SA SDC, sodium alginate soft dendritic colloid.

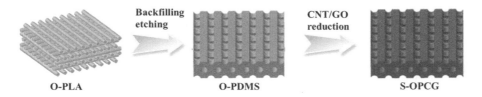

FIGURE 8.14 Schematic illustration of S-OPCG preparation.

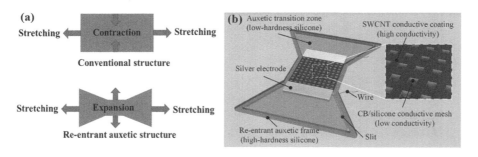

FIGURE 8.15 Auxetic strain sensor. (a) Deformation schematic diagram before and after stretching obtained from FEA simulation: conventional structure (top) and re-entrant auxetic structure (bottom). (b) Illustration diagram of the auxetic strain sensor. FEA, finite element analysis.

using DIW 3D printing and ink spraying technology proposed and prepared a novel auxiliary double conductive mesh strain sensor (ABSS) composed of multi-hardness silicone. Double conductive sheets were coated on stretchable CB-doped Ecoflex silicone rubber (CB/Ecoflex) sheets by a thin layer of highly conductive and crackable single-walled carbon nanotubes.

3D-Printed Flexible Strain Sensors

8.4.2 Bionic Structure

Inspired by threads prepared from spidroin with discrepant mechanical properties, Liu et al. [33] designed a graphene tactile sensor combined with dense filling and porous graphene microstructure (Figure 8.16). The porous structure of graphene has good electrical conductivity and can transmit sensor signals smoothly. This feature of division of labor and cooperation makes the sensor have multi-resolution positioning, position motion tracking, and tactile sensing and control functions, which has a broad application prospect in the multi-functional integration of electronic devices in the future. Based on the principle of bionic skin and hair, Xiang et al. [34] fabricated a sensor with a custom cross-lever structure to detect tensile and out-of-plane forces (Figure 8.17). The synergistic effect of the appropriate CNTs-to-GNPs mass ratio

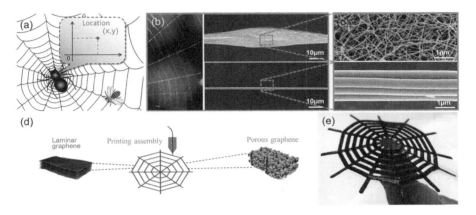

FIGURE 8.16 Spider-web–inspired multi-resolution graphene tactile sensor. (a) The signal transmission and location discrimination of natural spider web. (b) Optical photograph and SEM images of natural spider web: spiral threads (upper) and radial threads (lower). (c) Magnified SEM images of spiral threads (upper) and radial threads (lower). (d) Schematic illustration of printing manipulated graphene nanosheets assembled into different microstructures. (e) Optical photograph of the multi-resolution graphene tactile sensor under pressure.

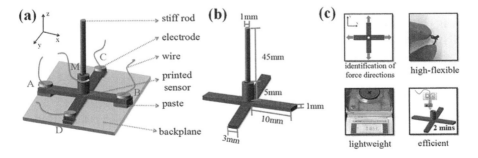

FIGURE 8.17 (a) Assembly schemes and (b) dimensions of the force sensor. (c) Characteristics of the printed sensor.

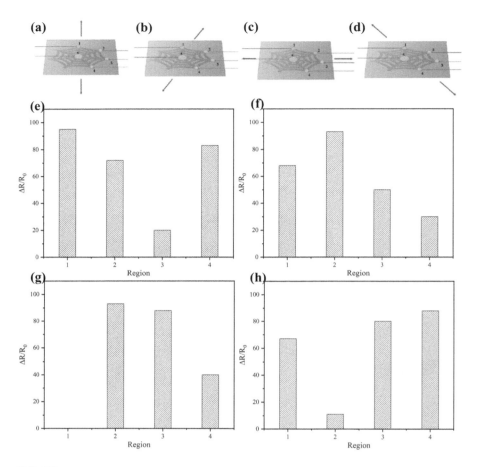

FIGURE 8.18 Schematic diagram of the sensor under 10% strain applied in different tensile angles: (a) 0°, (b) 45°, (c) 90°, and (d) 135°. The arrows show the tensile directions; The $\Delta R/R_0$ of the strain sensor at different tensile angles at 10% strain: (e) 0°, (f) 45°, (g) 95°, and (h) 135°.

was shown to improve the sensing performance of the sensor. From Figure 8.18, Chen et al. [35] designed and prepared a spider web-like strain sensor based on MWCNT/GNP/TPU CPCs. In addition, due to the special structural design, the sensor identified external forces in different directions.

8.4.3 MICROSTRUCTURE CHANNELS

3D printing is a rapid prototyping manufacturing technology, which is famous for its high design freedom and geometric complexity. 3D printing technology stands out for its low cost, high construction rate, resolution, precision of parts, and wide range of materials. Su et al. [36] proposed a novel selective wearable sensor with bending/stretching force differential and excellent signal performance (Figure 8.19). NaCl-doped agarose gel (NaCl@AG) was used as biocompatible conductive filler, and 3D printing elastomer shaper was used as the supporting matrix of wearable sensor.

3D-Printed Flexible Strain Sensors

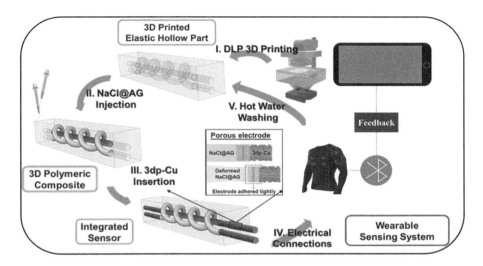

FIGURE 8.19 Schematic of the fabrication of 3D integrated polymeric wearable sensor.

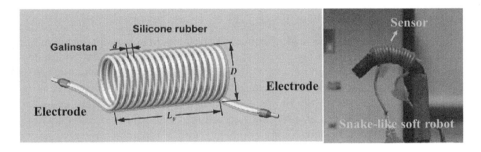

FIGURE 8.20 Illustration of the flexible multifunction inductance sensor.

AG has an interesting sol-gel transition property. NaCl@AG (sol) can uniformly form various three-dimensional structures through a three-dimensional printing microchannel elastomer forming device. As shown in Figure 8.20, Zhou et al. [37] described a novel 3D printing multifunctional inductance flexible and scalable sensor, which can measure axial tension and curvature. The sensor is made of silicone rubber and liquid metals coaxially printed. Due to the shape of the solenoid valve, the sensor can be easily installed on the snake-like soft robot and can accurately distinguish different degrees of tensile and bending deformation.

8.5 APPLICATION OF 3D-PRINTED STRAIN SENSORS

8.5.1 Electronic Skin

Electronic skin is a new type of electronic device that simulates human skin to feel external stimuli (pressure, temperature, humidity) through the integration and feedback of electrical signals. As a kind of flexible tactile bionic sensor, electronic skin

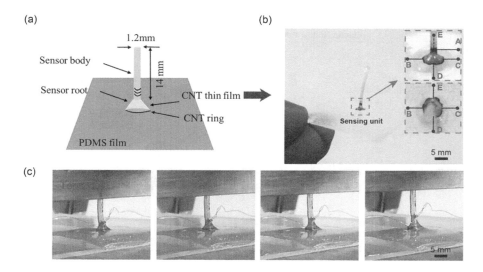

FIGURE 8.21 (a and b) 3D-structured stretchable strain sensors with out-of-plane multirings by the 3D self-pinning effect and (c) their performance with respect to strain sensing.

has been widely used in the field of human physiological parameters detection and robot tactile perception. In recent years, electronic skin has attracted wide attention from researchers around the world. The traditional electronic skin tactile sensor based on metal and semiconductor materials has been difficult to meet the requirements of tensile and portability in practical use due to its poor flexibility and wearability. Figure 8.21, Liu et al. [38] reported a three-dimensional stretchable strain sensor using 3D printing technology combined with non-planar capillary-assisted CNT self-pinning technology to monitor out-of-plane forces. It can be easily integrated and used to detect multi-strain, tiny gas, and fluid flows. Complex flow details can also be monitored, including velocity, damping vibration, quasi-static equilibrium, and flow state.

8.5.2 Soft Robotic Systems

Traditional robot system is composed of rigid bodies, actuators, and sensors. Unfortunately, many of these developed actuators and sensors cannot be transferred to software. Therefore, researchers engaged in soft robot research need to re-invent actuators and sensors for soft moving objects. The design of soft actuators and sensors must start with material selection and composition, as they are the basis around the actuators and sensors. From Figure 8.22, Truby et al. [39] reported a method of creating software-sensitive actuators (SSAs) by embedded 3D printing. This method is dominated by a variety of conductive features, while achieving tactile, ontological, and thermal sensations. This novel fabrication method can seamlessly integrate various ionic conductivity and fluid properties into the elastomer matrix to produce SSAs with desired biomimetic sensing and driving capabilities.

3D-Printed Flexible Strain Sensors 157

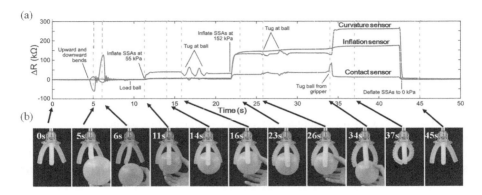

FIGURE 8.22 Soft robotic grippers with somatosensory feedback. (a) ΔR of each sensor as a function of time during the interaction sequence. (b) Images of an interaction sequence between a ball and a soft robotic gripper comprised SSAs. SSAs, software-sensitive actuators.

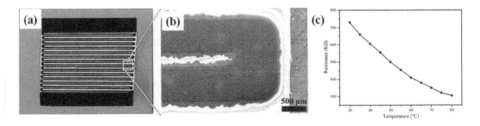

FIGURE 8.23 (a) Digital photograph of the TLTS device. (b) Zoomed image of TLTS device. (c) Relationship between the resistance R of the integrated electronic device and the temperature T. TLTS, linear temperature sensor.

8.5.3 Wearable Electronic Devices

Wearable technology is an emerging trend, which integrates cutting-edge technologies into daily activities. This technology conforms to the most changing and active way of life in this century. At present, wearable devices are mainly used for in vitro detection of vital signs and exercise health. Wearable sensors directly or indirectly contact with the human body, and real-time detection and transmission of some indicators (such as heartbeat, limb movement, and pulse) data. As shown in Figure 8.23, Zhao et al. [12] constructed an integrated electronic device by integrating 3D-printed asymmetric micro-supercapacitor and series linear temperature sensor (TLTS) based on rGO.

8.6 CONCLUSIONS

More attention has been paid to the design and preparation of flexible strain sensors using the characteristics of low cost, high precision, flexibility, and designability of 3D printing. This chapter reviews the composition, preparation methods, structural characteristics, and future application prospects of 3D printing strain sensors. In particular, we focus on some of our work in the preparation of flexible strain sensors using FDM method and introduce non-covalent bond modification or coordination effect between different dimensional direct nanofillers to further improve the overall performance of the sensor. Non-covalent modification of PCA significantly improved the dispersion of CNTs in the TPU matrix, which greatly improved the conductivity and sensing performance of the strain sensor. In addition, the conductivity and sensing performance of printed composites could be improved by using the coordination effect between nanofillers with different dimensions. More importantly, 3D printing technology with flexibility and designability can carry out complex design of the structure of the strain sensor, thus giving the sensor the ability to identify different types and different directions of strain. In summary, 3D-printed strain sensor shows broad application potential in wearable electronic equipment, human–computer interaction, intelligent robots, electronic skin, and other fields.

REFERENCES

1. Wallin TJ, Pikul J, Shepherd RF. 3D printing of soft robotic systems. *Nat Rev Mater.* 2018;3(6):84–100.
2. Zhao Z, Peng F, Cavicchi KA, Cakmak M, Weiss RA, Vogt BD. Three-dimensional printed shape memory objects based on an olefin ionomer of zinc-neutralized poly(ethylene-co-methacrylic acid). *ACS Appl Mater Interfaces.* 2017;9(32):27239–49.
3. Zhang D, Chi B, Li B, Gao Z, Du Y, Guo J, et al. Fabrication of highly conductive graphene flexible circuits by 3D printing. *Synth Met.* 2016;217:79–86.
4. Vaithilingam J, Simonelli M, Saleh E, Senin N, Wildman RD, Hague RJ, et al. Combined inkjet printing and infrared sintering of silver nanoparticles using a swathe-by-swathe and layer-by-layer approach for 3-dimensional structures. *ACS Appl Mater Interfaces.* 2017;9(7):6560–70.
5. Kim JY, Ji S, Jung S, Ryu BH, Kim HS, Lee SS, et al. 3D printable composite dough for stretchable, ultrasensitive and body-patchable strain sensors. *Nanoscale.* 2017;9(31):11035–46.
6. Goncalves J, Lima P, Krause B, Potschke P, Lafont U, Gomes JR, et al. Electrically conductive polyetheretherketone nanocomposite filaments: From production to fused deposition modeling. *Polymers (Basel).* 2018;10(8):925.
7. Fang L, Chen T, Li R, Liu S. Application of embedded fiber bragg grating (FBG) sensors in monitoring health to 3D printing structures. *IEEE Sens J.* 2016;16(17):6604–10.
8. Chen Q, Mangadlao JD, Wallat J, De Leon A, Pokorski JK, Advincula RC. 3D printing biocompatible polyurethane/poly(lactic acid)/graphene oxide nanocomposites: Anisotropic properties. *ACS Appl Mater Interfaces.* 2017;9(4):4015–23.
9. Liu H, Zhang H, Han W, Lin H, Li R, Zhu J, et al. 3D printed flexible strain sensors: From printing to devices and signals. *Adv Mater.* 2021;33(8):e2004782.
10. Muth JT, Vogt DM, Truby RL, Menguc Y, Kolesky DB, Wood RJ, et al. Embedded 3D printing of strain sensors within highly stretchable elastomers. *Adv Mater.* 2014;26(36):6307–12.
11. Valentine AD, Busbee TA, Boley JW, Raney JR, Chortos A, Kotikian A, et al. Hybrid 3D printing of soft electronics. *Adv Mater.* 2017;29(40): 1703817–24.
12. Pan W, Wallin T J, Odent J, et al. Optical stereolithography of antifouling zwitterionic hydrogels. J Mater Chem B, 2019;7(17):2855–2864.
13. Wei S, Zhang L, Li C, Tao S, Ding B, Zhu H, et al. Preparation of soft somatosensory-detecting materials via selective laser sintering. *J Mater Chem C.* 2019;7(22):6786–94.

14. Liu C, Huang N, Xu F, Tong J, Chen Z, Gui X, et al. 3D printing technologies for flexible tactile sensors toward wearable electronics and electronic skin. *Polymers (Basel)*. 2018;10(6):629–659.
15. Leigh SJ, Bradley RJ, Purssell CP, Billson DR, Hutchins DA. A simple, low-cost conductive composite material for 3D printing of electronic sensors. *PLoS One*. 2012;7(11):e49365.
16. Xiang D, Zhang X, Han Z, Zhang Z, Zhou Z, Harkin-Jones E, et al. 3D printed high-performance flexible strain sensors based on carbon nanotube and graphene nanoplatelet filled polymer composites. *Journal of Materials Science*. 2020;55(33):15769–86.
17. Xiang D, Zhang X, Li Y, Harkin-Jones E, Zheng Y, Wang L, et al. Enhanced performance of 3D printed highly elastic strain sensors of carbon nanotube/thermoplastic polyurethane nanocomposites via non-covalent interactions. *Compos Part B Eng*. 2019;176:107250.
18. Xiang D, Zhang X, Harkin-Jones E, Zhu W, Zhou Z, Shen Y, et al. Synergistic effects of hybrid conductive nanofillers on the performance of 3D printed highly elastic strain sensors. *Compos Part A Appl Sci Manuf*. 2020;129: 105730-42.
19. Ligon SC, Liska R, Stampfl J, Gurr M, Mulhaupt R. Polymers for 3D printing and customized additive manufacturing. *Chem Rev*. 2017;117(15):10212–90.
20. Xiang D, Liu L, Chen X, Wu Y, Wang M, Zhang J, et al. High-performance fiber strain sensor of carbon nanotube/thermoplastic polyurethane@styrene butadiene styrene with a double percolated structure. *Front Mater Sci*. 2022;16(1):99–110.
21. Xiang D, Zhang Z, Han Z, Zhang X, Zhou Z, Zhang J, et al. Effects of non-covalent interactions on the properties of 3D printed flexible piezoresistive strain sensors of conductive polymer composites. *Compos Interfaces*. 2020;28(6):577–91.
22. Zhang Z, Xiang D, Wu Y, Zhang J, Li Y, Wang M, et al. Effect of carbon black on the strain sensing property of 3D printed conductive polymer composites. *Appl Compos Mater*. 2022:1–14.
23. Zhang YZ, Wang Y, Jiang Q, El-Demellawi JK, Kim H, Alshareef HN. MXene printing and patterned coating for device applications. *Adv Mater*. 2020;32(21):e1908486.
24. Britton J, Krukiewicz K, Chandran M, Fernandez J, Poudel A, Sarasua JR, et al. A flexible strain-responsive sensor fabricated from a biocompatible electronic ink via an additive-manufacturing process. *Mater Des*. 2021;206: 109700-10.
25. Kwon SN, Kim SW, Kim IG, Hong YK, Na SI. Direct 3D printing of graphene nanoplatelet/silver nanoparticle-based nanocomposites for multiaxial piezoresistive sensor applications. *Adv Mater Technol*. 2018;4(2):1800500.
26. Cao WT, Ma C, Mao DS, Zhang J, Ma MG, Chen F. MXene-reinforced cellulose nanofibril inks for 3D-printed smart fibres and textiles. *Adv Funct Mater*. 2019;29(51):1905898–909.
27. Wei S, Qu G, Luo G, Huang Y, Zhang H, Zhou X, et al. Scalable and automated fabrication of conductive tough-hydrogel microfibers with ultrastretchability, 3D printability, and stress sensitivity. *ACS Appl Mater Interfaces*. 2018;10(13):11204–12.
28. Liu S, Li L. Ultrastretchable and self-healing double-network hydrogel for 3D printing and strain sensor. *ACS Appl Mater Interfaces*. 2017;9(31):26429–37.
29. Williams AH, Roh S, Jacob AR, Stoyanov SD, Hsiao L, Velev OD. Printable homocomposite hydrogels with synergistically reinforced molecular-colloidal networks. *Nat Commun*. 2021;12(1):2834.
30. Chen Q, Cao P-F, Advincula RC. Mechanically robust, ultraelastic hierarchical foam with tunable properties via 3D printing. *Adv Funct Mater*. 2018;28(21):1800631–39.
31. Duan S, Yang K, Wang Z, Chen M, Zhang L, Zhang H, et al. Fabrication of highly stretchable conductors based on 3D printed porous poly(dimethylsiloxane) and conductive carbon nanotubes/graphene network. *ACS Appl Mater Interfaces*. 2016;8(3):2187–92.

32. Wang Z, Luan C, Liao G, Liu J, Yao X, Fu J. High-performance auxetic bilayer conductive mesh-based multi-material integrated stretchable strain sensors. *ACS Appl Mater Interfaces*. 2021;13(19):23038–48.
33. Liu L, Huang Y, Li F, Ma Y, Li W, Su M, et al. Spider-web inspired multi-resolution graphene tactile sensor. *Chem Commun (Camb)*. 2018;54(38):4810–13.
34. Xiang D, Zhang Z, Wu Y, Shen J, Harkin-Jones E, Li Z, et al. 3D-printed flexible piezoresistive sensors for stretching and out-of-plane forces. *Macromol Mater Eng*. 2021;306(11):202100437–46.
35. Chen X, Zhang X, Xiang D, Wu Y, Zhao C, Li H, et al. 3D printed high-performance spider web-like flexible strain sensors with directional strain recognition based on conductive polymer composites. *Mater Lett*. 2022;306: 130935-8.
36. Su X, Borayek R, Li X, Herng TS, Tian D, Lim GJH, et al. Integrated wearable sensors with bending/stretching selectivity and extremely enhanced sensitivity derived from agarose-based ionic conductor and its 3D-shaping. *Chem Eng J*. 2020;389:124503–13.
37. Zhou LY, Gao Q, Zhan JF, Xie CQ, Fu JZ, He Y. Three-dimensional printed wearable sensors with liquid metals for detecting the pose of snakelike soft robots. *ACS Appl Mater Interfaces*. 2018;10(27):23208–17.
38. Liu Z, Qi D, Leow WR, Yu J, Xiloyannnis M, Cappello L, et al. 3D-structured stretchable strain sensors for out-of-plane force detection. *Adv Mater*. 2018;30(26):e1707285.
39. Truby RL, Wehner M, Grosskopf AK, Vogt DM, Uzel SGM, Wood RJ, et al. Soft somatosensitive actuators via embedded 3D printing. *Adv Mater*. 2018;30(15):e1706383.

Index

biaxial stretching 3, 43–44
blown film extrusion 3, 77–78

carbonization 131–32
chemical vapor deposition 9, 129
compression molding 3, 20
conductive polymer composites
    preparation 2–3
    processing 3–4
    properties 4–6

damage self-sensing 103–08
double percolated structure 7, 98–100

electronic skin 136, 155–56

freeze-drying 129–31

human-computer interaction 121–22, 136

layer-by-layer coating 114–15

robotics 122, 156–57

segregated structure 7, 98–100, 105, 107
spinning 117

3D printing 9, 141–45

volume resistivity 26–29, 32–33, 37–38, 53–55, 62–63, 69–70, 88–90

wearable electronic device 135–36, 157

Xiang, D. 7, 83, 99, 100, 104, 144–46, 153

161

Taylor & Francis eBooks

www.taylorfrancis.com

A single destination for eBooks from Taylor & Francis with increased functionality and an improved user experience to meet the needs of our customers.

90,000+ eBooks of award-winning academic content in Humanities, Social Science, Science, Technology, Engineering, and Medical written by a global network of editors and authors.

TAYLOR & FRANCIS EBOOKS OFFERS:

- A streamlined experience for our library customers
- A single point of discovery for all of our eBook content
- Improved search and discovery of content at both book and chapter level

REQUEST A FREE TRIAL
support@taylorfrancis.com